A PRIMER OF GIS

A Primer of GIS

Fundamental Geographic and Cartographic Concepts

SECOND EDITION

Francis Harvey

THE GUILFORD PRESS
New York   London

Library of Congress Cataloging-in-Publication Data

Harvey, Francis (Francis James)
    A primer of GIS : fundamental geographic and cartographic concepts /
Francis Harvey. — Second edition.
        pages cm
    Includes bibliographical references and index.
    ISBN 978-1-4625-2217-0 (pbk.) — ISBN 978-1-4625-2218-7 (hardcover)
    1. Geographic information systems.   2. Cartography.   I. Title.
    G70.212.H38 2016
    910.285—dc23
                                                                    2015012170

For Anna and Alicja
Thanks again for everything.

Preface

Welcome to the world of cartography, geographic information (GI), geographic information systems (GIS), and the second edition of *A Primer of GIS*!

The idea behind this book is simple: to put in the hands of people interested in GIS, geographic information science, and geospatial science and engineering a book that provides a broad preparation for later work with the geographic information that underpins GIS. It covers what and how GI represents, analyzes, and communicates about human and environmental activities and events on our planet.

The book's uses are broad: some people read this as a cartography book, for others it serves as a basic introduction to GIS, others use it in self-guided study—this range of uses underscores the diversity of GIS and the relevance of the book's approach. Higher education instructors in a variety of programs—traditional GIS, domain sciences, technical training, geography undergraduate programs, and GIScience curricula—have used the book, which serves a broad range of pedagogical needs and outcomes in GIS use and training. *A Primer of GIS* remains unique in the field because it is not software-focused. Rather, this book provides a conceptual introduction to the principles behind GIS without requiring the use of GIS software. Through practical examples and exercises, you will find in this book detailed introductions to the theories, concepts, and skills you will need to prepare for working with GIS, regardless of your educational background or interests.

Though the major learning outcomes I pursue in this book remain unchanged, it has been substantially revised and updated from the first edition, based on suggestions from adopters of the first edition, comments from readers, and detailed remarks from reviewers of the draft of the manuscript. Their comments, insights, and helpful suggestions aided me immensely in preparing the book.

Organization

Reflecting the successful organization of the first edition, the second edition keeps the hybrid hierarchical/linear division of parts and chapters to ensure that the book supports a broad range of instructional objectives: merging fundamental principles of geography, demonstrating elementary techniques for geographic data collection and cartographic representation, and introducing geographical analysis. The modular approach allows for use of the book at different levels, or as a phased introduction that deepens student comprehension of GIS fundamentals. The book's organization provides a clear structure for many curricula and courses. The first part provides an overview of basic concepts, fundamentals, and a review of GIS uses. The second part digs into fundamentals and functions. The third part goes into techniques and practices of representation and communication. The fourth part introduces key GI analysis approaches and looks to the past and future. The parts can be taught in sequence or individually; specific chapters can be used to further suit your pedagogical and curricular requirements. A brief introduction to each part provides a concise overview of the chapters in that section.

Throughout the book I present GIS topics to students from the point of view of GIS developers—what resources does the GIS developer draw on to shape GI into a GIS? In Part 1 I present a conception of GIS in which it is evaluated in terms of how well it represents and communicates. If we are ever fooled into thinking that the world *is* the way a map represents it, cartography has succeeded. The tools cartography has developed to perform this wonderful deception are the subject of the chapters in Part 2: projections, coordinate systems, databases, surveying, remote sensing, and various locational systems that go beyond simple *x, y* coordinates. Part 3 takes up cartographic representation in its social context: cultural aspects, propaganda, privacy issues of GPS, how we organize space governmentally and/or administratively, and the basics of web and Internet mapping. With the proliferation of sources of data, and the ease of updating and correcting data due to GPS and Internet interactivity, our ability to analyze GI has grown substantially. Part 4 looks at how GI is analyzed, and the last chapter of Part 4, and of the book, is an overview of the history of GIS and what to look forward to in the future.

The Learning Outcomes for This Book

The major learning outcomes are that students should (1) understand the principles of cartography and geography that are at work in GIS and in GIScience; (2) apply these principles in practical examples and exercises, which are at the end of each chapter; and (3) understand the choices made in producing maps and GI, which are always a selective representation of the world.

Changes from the First Edition

All the chapters underwent revision and two new chapters were prompted by changes that have taken place in GIS since the publication of the previous edition. The new chapters cover uses of GIS that have sprouted up in the last decade (Chapter 4), and online mapping and Big Data (Chapter 14). Particular areas of the field that receive enhanced coverage in this edition include:

- The increase in online mapping
- Uses of geocoding
- The issue of locational privacy in GPS and security cameras
- Cartographic principles for online mapping
- The meaning of representation
- Distinctions among data types
- Global navigation satellite systems
- Public participatory GIS
- Vector and raster data models
- Datums
- Remote sensing
- The representation of space and time as things or events
- Projections

Though the organization of the book is in most ways similar to the first edition, the second includes a glossary and section overviews with graphics and text to help orient instructors and students.

Pedagogical Features and Resources

This book includes a number of features to enhance students' learning of the material. Access point boxes in some chapters provide detailed practical examples of how people use GI; example boxes focus on relevant aspects of the textual discussion; and In Depth sidebars offer a closer look at theories and concepts. To assist your reading and learning, you will find at the end of each chapter review questions concerning the chapter's material and exercises to put your knowledge into practice. These include Extended Exercises that, when followed, prompt a considerable depth of involvement in the topic. End-of-chapter materials also include a useful range of articles, books, and Internet links relevant to your interests or learning needs. Links to Internet sites I recommend can be found on the companion website to this book at *www.guilford.com/harvey2-materials*. The glossary, new to this edition, helps tie together many of the concepts introduced.

Acknowledgments

Numerous students, colleagues, and friends have discussed with me the broad range of topics that are covered by this book. I can't mention one without mentioning all, so thanks to all of you.

I would like to thank the following individual reviewers of the book, whose many comments, insights, and helpful suggestions aided me immensely in guiding the large and small decisions regarding the revision: Jeff Hamerlinck, Director, Wyoming Geographic Information Science Center, University of Wyoming; Werner Kuhn, Jack and Laura Dangermond Chair, Department of Geography, University of California, Santa Barbara; Jun Liang, Department of Geography, University of North Carolina at Chapel Hill; Timothy LeDoux, Geography Department, Westfield State University; Patrick McHaffie, Department of Geography, DePaul University; Karen E. Blevins, Geography Department, Mesa Community College; and Holly M. Widen, PhD candidate, Department of Geography, Florida State University.

Several instructors who used the first edition were extremely generous with their time and explained to me their use of the book in the classroom and their experience of its strengths and limitations. To them I owe special thanks: Adam Iwaniak, Marek Baranowski, Brett Black, Amir Chaudhry, Nathan Clough, Allen Lin, Randy Johnson, Chris Lloyd, William Mackaness, Annamaria Orla-Bukowska, and Nick Tate. To Lars Hellvig I owe special thanks for his careful reading and suggestion of clarifications to the discussion of metric scales. I would also like to thank a colleague at the University of Minnesota, Mark Lindberg, for his considerable help with the figures. I want to thank the professionals at The Guilford Press who have helped me through the many stages of preparing this second edition, especially the thorough and thoughtful William Meyer and Oliver Sharpe, who worked on the redesigned interior of the second edition. Most of all I want to thank Alicja Piasecka and Anna Piasecka for their support during the many hours spent revising this book. Big and small changes made in the revision attempt to respond to the many constructive suggestions and reports of the varied uses of *A Primer of GIS* I've received. I hope the suggestions have been implemented well and that second edition is a worthwhile revision and update.

Without the help of the people named above and the unnamed I could not have written this book; nevertheless, misinterpretations or errors in the presentations or translations are my sole responsibility.

I appreciate the comments and reviews of readers, now and in the future. Do you find this edition serves your needs? How can the next edition, possibly in a "post-GIS" era, continue to keep up? For these or any other *A Primer of GIS*–related questions, please email me at *francis.harvey@gmail.com*.

Contents

CHAPTER 3

GI and Cartography Issues 54

CHAPTER 4

Some History and the Continued Many Uses of GIS 73

PART 2

Fundamentals and Functions 83

CHAPTER 5

Projections 85

CHAPTER 6

Locational and Coordinate Systems

CHAPTER 7

Databases, Cartography, and GI

CHAPTER 8

Surveying, GPS, Digitization

CHAPTER 9

Remote Sensing

CHAPTER 10

Locations and Fields: Discrete and Nondiscrete GI

PART 3

Techniques and Practices

CHAPTER 11

Cartographic Representation

CHAPTER 16

CHAPTER 17

Downloadable PowerPoint slides of all original figures, photos,
and tables, as well as the Web Resources, are available at
www.guilford.com/harvey2-materials.

PART 1

Your World and GI Technology

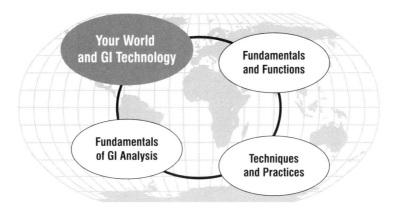

This section of the book, Your World and Geographic Information Technology, provides a general introduction to fundamental concepts from cartography and geographic information (GI). Detailed discussions of these concepts follow in other parts and chapters. Depending on your background and knowledge, some of this material may proverbially be part of a road you know well. Even if some of this section seems introductory in nature or even at times redundant, curious readers with previous GI, cartography, or related experiences still may want to skim these chapters to get an overview, refresh knowledge, and get a better sense of how this book approaches fundamental concepts and connects the challenges of representing the world around us to fundamentals, functions, techniques, and practices. The overview characteristic of the chapters in this section inexorably means some readers will encounter overlaps with chapters in later sections of *A Primer of GIS*. In summary, Chapter 1 provides a very general overview of basic concepts and an introduction to the book. Chapter 2 uses *choice* as a way of thinking about how we create representations of geographical subjects so that what we communicate is reliable and accurate. When making choices, the most basic or fundamental of these are covered in Chapter 3. Chapter 4 begins to look more specifically at GIS, and provides a brief history of the use of GIS in problem solving with some examples of what GIS does, again using the choices made in setting up GIS as a way into the topic. The later chapters, in Parts 2, 3, and 4, will return to the fundamentals addressed here in Part 1, but in greater detail.

CHAPTER 1

Goals of Cartography and GI: Representation and Communication

Many of our representations and communications about things and events around us, in history, even in the future, rely on geography and cartography. Geographic information systems (GIS) too. Usually we simply forget how commonplace maps and geographic information (GI) are, so maybe you have never given them much thought. Nevertheless, maps and GI are essential to how we know the world. We regularly refer to maps to verify what we know, expand our knowledge, and get to know things we otherwise might never know. The endless complexity of the world around us presents us with a multitude of choices about what to represent and how to represent that complexity in the form of maps and as GI.

Right now, take a look out a window. If you have a map of the same area, also look at that map. Compare your view to the map or to a map you remember of the place you are looking at. They are obviously different. Try to make a list of the differences. What is different between the view and the map? There are many, many differences: trees, buildings, or sidewalks may be missing on the map, the color of the road on the map may set it apart from other roads, the connections between roads may be much plainer on the map than what you can see. How and why GI and maps are different from our experiences and observations are important questions that this book will help you understand. GIS involve many issues and choices. This chapter and the following three chapters provide an overview of some general issues and choices, with more detail to come in the other chapters of the book. As you read this now and look at the map and out the window at the same area, you can start thinking about how your observations and perceptions of things outside are different from those of the map: some things are missing, some things are simplified, and some things are exaggerated on a map. GI and maps are representations that follow a number of principles and conventions that help deal with the complexity of the world and guide choices that lead to clear communication. Should the map include sidewalks? Will the GI describe the height of buildings? Are trees distinguished by species? These choices also will determine the way locations on the spherical earth are transformed to a two-dimensional plane, the types of colors and symbols to use, and the types of questions that people will turn to the map or GI to help find answers for.

Consider two other examples that highlight the different types of representation used in maps and GI (i.e., the data stored on a computer that contains information for making maps or conducting analysis) and point to some of the principles and conventions that guide mapping choices. First are maps of continents or subcontinents. You probably never actually have seen the entire United States, all of Europe, or all of southern Asia in person, but you know something about how they are geographically arranged through maps (see Figure 1.1). Second, consider maps you use to help you get around

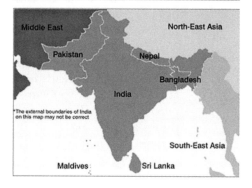

FIGURE 1.1. Three modern maps showing geology, landforms, and political boundaries; each relies on different forms of geographic representation and cartographic representation to communicate particular meanings. Representational concepts and conventions of color and scale are crucial to assuring that each map successfully communicates with its intended audience.

the place where you live. You may know the way to go when you travel to work or school partly from descriptions prepared with the help of geographers and maps made by cartographers. Starting with these two examples, if you pause to think about the many different uses and roles of geography and cartography in the last 500 years (an arbitrary period), starting with the European period of exploration and colonization, we can conclude that geographers and cartographers have helped people to understand, navigate, control, and govern most of our world for millennia (see Figure 1.2). Your world and the whole world would be much different without geography and cartography. We rely on these representations and the principles and conventions behind them to make sense out of the world in many different ways—sometimes GI and maps may be the only way to know something, other times they are important complements to other things we know or can ask. **Principles** are standard procedures that people in a field follow—for example, when a cartographer chooses a projection to make a map. **Conventions** are uses or procedures agreed upon by experts, but usually they have become common knowledge—for example, that north is the direction oriented at the top of a map. Sometimes we are sure about how things are geographically organized, but sometimes we may be less certain. We probably know where the city we live in lies in relationship to the coastline, but we may be less sure about whether New York or Boston is further to the east in the United States. A good representation takes these issues into account to assure that its readers or users find the representation helpful in communication (see Table 1.1).

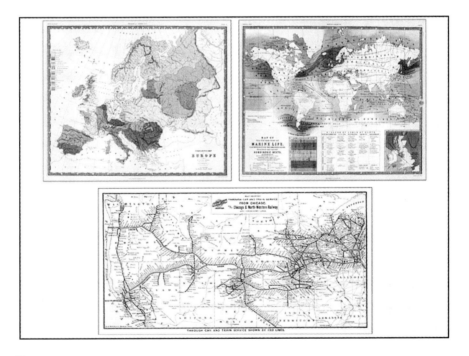

FIGURE 1.2. Three thematic maps from the 19th century that demonstrate different geographic representations and cartographic representations of things and events.

TABLE 1.1. Some Common Things and Their Representations

	Geographic representation (basic)	Cartographic representation
Stream	Line	Color blue
Road	Line (usually)	Color black or red
Forest	Polygon	Color green
Industry	Polygon	Color gray
County or district	Polygon	Dashed boundary line
Well	Point	Circle with cross
Land parcel	Polygon	Thin black boundary line
House	Polygon	Thick black outline
Lake	Polygon	Color blue
Park	Polygon	Color green
Sand dunes	Polygon	Black dots on sand-colored background

Modern geography and cartography share many principles and conventions that form symbiotic relationships, which make up an important basis for the geographic representation of the world in other scientific and professional fields. We define them in this book as follows. **Geography** analyzes and explains human and environmental phenomena and processes taking place on the earth's surface, thereby improving our understanding of the world. **Cartography** develops the theories, concepts, and skills for describing and visualizing the things and events or patterns and processes from geography and communicating this understanding. In this book *things* refer to elements of the world that are static, either by their nature or by definition. *Events* refer to selected moments in a process (see Figure 1.3). Both are representations involving our innate cognitive capabilities and culturally and socially influenced knowledge of the world. What geography analyzes and explains, cartography communicates visually. Geography and cartography are dynamic subjects that involve a broad set of theories, concepts, and skills that undergo constant development and refinement as knowledge, culture, and technology change. Because of their usefulness, geography and cartography are parts of many other human activities and disciplines. Biologists, geneticists, architects, planners, advertisers, soldiers, and doctors are just a few of the scientists and professionals who use geography and cartography. However, because geography and cartography are so commonplace, they are often easy to overlook. If you want to understand how to use and communicate better with maps, then you need to examine them closely and understand how and why GI and maps are different from what you see and observe. With a greater understanding of geography's and cartography's principles, conventions, and underlying basic concepts, you will be able to work better in any field.

For most people, maps are the most common way to learn about geography. But GI is very significant and continues to gain in importance. Geography and cartography have always been interdisciplinary fields. Many other

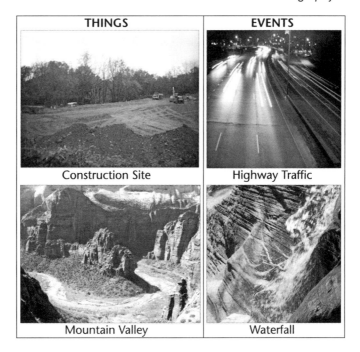

THINGS	EVENTS
Construction Site	Highway Traffic
Mountain Valley	Waterfall

FIGURE 1.3. GI and maps show things and events from built and natural environments. The primary difference is change. Things are static for the observer, whereas events record selected moments of a process.

disciplines and fields of human endeavor have drawn on their knowledge and skills and continue to do so. Recent technological innovations further broaden possibilities for people to make measurements of geographic things and events, operate and transform these measurements, and represent the measurements as information and maps. They produce geographic information, which is very easy to copy between computers, but often very hard to get out of the hands of the people and organizations who are responsible for that GI. Certainly, the circle of people working with concepts from geography and cartography has grown tremendously in the last 20 years. This has much to do with the increased availability of computers and programs for working with digital GI. That term sounds simple, but turns out to be highly complex. You might want to think about GI as you would about oxygen: you can't necessarily see it, but its presence has positive effects for people. Maps rely on GI. GI is, of course, very different from maps in many ways. One of the most fundamental differences is that GI is very, very easy to change, whereas maps, if changed, are usually somehow destroyed. That is certainly one of many reasons for the growth of online mapping in the last 10 years. It also means that GI can be used many times, which gives it a great advantage over maps.

Indeed, many geographers and cartographers would claim that GI makes geography and cartography more accessible than ever before. Farmers use global positioning system (GPS) technologies and satellite images to help disperse fertilizers and pesticides more accurately, safely, and economically. Fire departments route fire trucks to their destinations based on analysis of road networks and real-time traffic information. Millions of people use GPS when navigating a boat, flying a plane, or driving a car. Many cars now come equipped with satellite navigation systems that rely on dashboard map displays to help drivers find their way. GIS is used also in many research facilities and offices to help analyze and manage resources. And the literal explosion of data that is easily accessible through computer networks and the Internet makes it possible for more and more people to create and use GI. Improved geographic and cartographic technology has played a key part in important economic developments not only now, but in the past as well. The astrolabe used by navigators in the Middle Ages changed the way locations were determined and mapped; exploration consequently became more accurate and safer. Offset printing, introduced in the late 19th century, made it possible to produce series of maps by using combinations of different plates; maps then became commonplace in books, magazines, and newspapers. The most significant current geographic and cartographic innovations arise from the computer and the development of information technology for processing data during the last 40 years. The fields of geography and cartography entered an unparalleled period of symbiosis with the introduction of information technology for processing GI. This symbiosis resulted in a new field called geographic information systems (GIS), which, since the 1960s work by Roger Tomlinson, Edgar Horwood, William Warntz, and many others, has grown into a major information technology field and a science.

People from many academic backgrounds correctly point out that the relationship between geography and cartography has changed and continues to change as a result of technological change; sometimes they even question the future of cartography because of GIS. Now, some people assume, computers can do all cartography, and access to data means we've entered an era freed from the specialized knowledge of geographers and cartographers. However, it is apparent that many of the key geographic and cartographic concepts established over thousands of years remain important. In fact, one could claim that these fields are really not changing conceptually, but only in degrees. As information technology becomes commonplace, many more people are now able to do things without the years of training that only cartographers and geographers previously had. However, making a map and making a *good* map are often different things. Online maps and neogeographic concepts are new and have much potential, but they still rely on GI, and the underlying concepts of geography and cartography remain fundamental. Indeed, because of all the people now doing work with geography and cartography on computers, one could also argue today that the underlying concepts and skills of geography and cartography have become more relevant for more pursuits and sciences. Both are certainly true; however, without understanding of key concepts and skills, the best intentions can easily go wrong. Obviously, professionals always need to produce the highest quality maps and always benefit from better understanding of the

IN DEPTH Globes

For many people globes seem to be ideal cartographic representations of the earth's geography. They are certainly attractive, but for a number of reasons they are limited in their use and suitability for most maps and geographic information. They remain, however, the best reference for understanding the earth's three-dimensional shape and for conceptualizing latitude and longitude.

Making a Globe

A globe ends up round, but it is printed on a flat piece of paper just like any other map. The map is divided, and later cut, into what is called a "gore." When the paper is glued to the round base, these strips fit together, resulting in an uninterrupted sphere. This technique has been used for several hundred years.

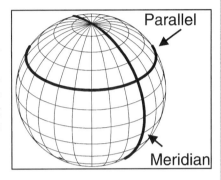

Globes indicate lines of latitude and longitude, also called parallels and meridians, respectively.

Reading a Globe

Locations on a globe can be found by latitude and longitude. **Latitude** is a degree measure used to indicate the spherical distance from the equator; parallels run from east to west at a constant latitude.

Longitude is another degree measure that has a fixed origin, nowadays the Greenwich meridian, running through Greenwich, England, in southeast London, for indicating east/west location from the origin. Meridians run from pole to pole.

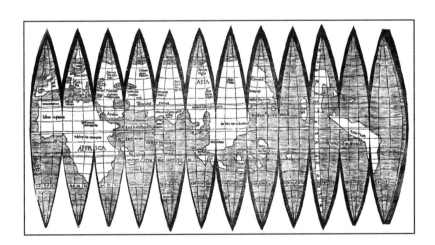

Waldseemüller globe gore from 1507. This would be cut out and pasted on a sphere to create a globe.

concepts and skills—regardless of what and how much information technology is capable of doing.

Representation and Communication

In this book, you will learn about both the old concepts and the new concepts within cartography and geography. You will find that the old and the new concepts of geography merge with new information technologies in representation and communication, the two essential activities of geography and cartography. In this book **representation** refers to the active process of observing the world and symbolizing those observations to make meaning. **Communication** means the process of presenting these representations by some people and the viewing, or reading, of those representations by other people. Geographers and cartographers are always involved in communication, for even if it is not their immediate goal, maps and GI are always made to share information and knowledge about the world. A **geographic representation** is the specific process of abstracting observations of the world into things or events, often resulting in a model. **Things** are static. They are often the result of human activities or they are measurements of the properties of objects or features. An **event** records the state of a process at a particular moment of the process—for example, the movement of cars and trucks, the flow of water, the melting of ice, or the spread of a disease. A **cartographic representation** involves the process of symbolizing the geographic representation. Successfully communicating information about

---

IN DEPTH **Representation**

An artist may represent a face, a room, or a landscape. A computer scientist may represent the same things and events as a data model or a database. A geographer may represent these things as objects and as relationships. A cartographer may represent them as features and attributes.

For some people, representations have dangerously replaced experience. On the other end of the spectrum, other people find that our enhanced technologies for representation make it possible to transcend past limitations. We can talk to people all over the world, they may point out, but others ask, For what good? This is not the place to engage these discussions, but before getting into more specifics about representation, it's good to have some notion of how the term *representation* is used.

All these forms of artistic, computer, geographic, and cartographic representation are important to how we make sense of the world around us, but our vocabulary is a bit weak for distinguishing between them. To help with this problem, this book always precedes representation with an adjective to distinguish what kind of representation is meant—for example, geographic representation.

things and events requires you to know something about geographic and cartographic representation. These two concepts include color, symbology, modeling, projections, and, now with GIS, spatial database queries and attribute types (all covered in later chapters).

This book considers representation and communication as related and fundamental topics in geography and cartography. A peculiar geographic fascination is common among people working with GIS, whether they work for a utility company, a county government, a university administration, or a corporate marketing department: How can the infinite complexity of the earth's surface and related processes be reliably represented? This seemingly abstract question touches on the key issues these people have become aware of through their education, training, and work experiences. They must decide how to represent selected things as patterns that show important elements and processes in relation to the places where they take place. Figure 1.4 simplistically shows a few basic choices and the different ways events can be represented either by highlighting the process or by translating the site of the process into a pattern. How representation is chosen also must consider the context of the intended communication, particularly the reader's/user's knowledge and background: How well does the application or map correspond to what the readers/users know or could know? Are data available to provide that information? How long would it take to acquire new data? The issues include many specific questions—for example, Is it sufficient to show trees as points where their trunks are located or as areas that show the reach of the foliage?

The answers to the question of representation usually come down to choices and quality. There are many choices: How much detail is needed for

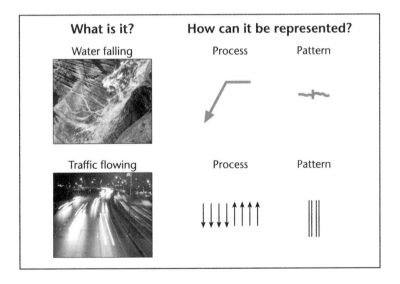

FIGURE 1.4. Events can be represented in maps both as representations of processes and as the resulting patterns.

the GI or map? Will a poster-size paper map be needed for the detail or does it have to fit on a small handout? How accurate should it be? How big should (or can) it be? If this is a computer application, will the data be available on a CD-ROM or a DVD, or will it be downloadable over the Internet? How big is the screen display? The type of communication and the background of potential users also need to be taken into consideration: Will specialists use the map or application? How much knowledge do they have about the area? What is/are the specialists' purpose or purposes? How much contextual information is required? How abstract can the representation be? How reliable must the representation be? Each decision influences quality in complex ways. If the map needs to fit on a small piece of paper, but the area of an entire state or province needs to be shown, it will be very difficult to show a great deal of detail.

Issues related to wise choices and quality come back to perennial issues for geographic representation. The space of the earth's surface is limited, but because all GI is an abstraction with no limit to the number of choices we may make in presenting it, geography's potential representations are unlimited. The space of the earth's surface shows itself in peculiar characteristics in every representation. How close objects are on paper or on the screen depends on the relationship between the size of the representation and the actual area on the ground. This is what geographers refer to as **scale**. Scale is a crucial component of geographic representation and cartographic communication. Of course, people think objects closer together on a map are more related to each other than objects far apart, but if you consider the scale of the representation, even close objects may actually be very distant from one another. The issue of scale is a particularly central concept for all GI because of the ways it allows and restricts representation, the communication of relationships between things, and the interpretation of geographic associations.

The Power of Maps

Successful answers to these questions and attention to the decisions made in representing geography are what gives GI power and makes maps powerful, to borrow from Denis Wood's thoughtful writing about maps. A map or GI application is selective and greatly limited, but it remains a key means of understanding and analyzing the world. This power is very attractive and lucrative; its misuse and abuse lend support to many ill-conceived projects (see Figure 1.5).

Maps are powerful for a number of reasons. But perhaps the most elementary reason is that they offer an authoritative representation of things and events in the world that we cannot otherwise experience in a single moment. Most maps, even the most mundane kind of map—for example, one showing temperatures across North America—show us things, events, and relationships that you could never experience yourself in a similar

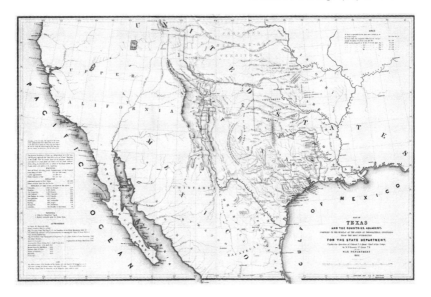

FIGURE 1.5. The power of maps often depends on the currency of the map. In 1844, when this map was prepared for the U.S. State Department, it played an important role in helping people understand the Texas conflict.

complete but quickly grasped form. You can read a book, look at a photograph, watch a film, check things out on the Web, but a successful map easily and quietly combines much detail into a synoptic representation (see Figure 1.6).

Perhaps the second elementary reason why maps are so powerful is that they represent something beyond our own limited experience: other people and places we may never see in person, other things or events that we may never know about otherwise. They became a key source of information about people and places we can't experience because of distance or because of complexity. Maps become a primary source of information for many things since we often cannot verify what they tell us. Is the Eiffel Tower located at the center of Paris? Unless you are in Paris or will be shortly the answer to that question cannot be determined except by using a map.

The power of maps comes through their ability to create representations of the world that most people won't question because they lack the direct experience of the people, places, things, or events to evaluate the representations. It is very hard to know that a representation implicitly makes a threat out of a neighbor, errs in creating symbols that mask important details, or explicitly shows a part of the world in a biased manner. For example, using red to show the country of one's enemy awakens a sense of menace because most Westerners associate the color red with danger. Showing a country in green has the opposite impact. Because they follow frameworks and conventions that we have become used to, slight distortions are easily veiled and become undistinguishable.

FIGURE 1.6. The power of maps is significant for associating organizations with a nation or region. This sign for the Polish Tourist Association uses an iconic representation of Poland's national boundaries. It assures people associate its services with travel in Poland.

Maps are often misused and have become important tools for propaganda and advertising (see Chapter 12). Extreme examples clearly show abuse of cartographic integrity, but you also need to be wary of more common and subtle misuse of map power to create biased representations.

Types of Maps

Three of the most common types of maps are thematic, topographic, and cadastral (see Figure 1.7). There are many ways to develop typologies of maps, but these three types seem to distinguish both how and why maps are used. **Thematic maps** are the most common: they show specific topics and their geographic relationships and distributions. Thematic maps show us the weather forecast, election results, poverty, soil types, and the spread of a virus. **Topographic maps**—from the United States Geological Survey (USGS), for example—show the physical characteristics of land in an area and the built changes in the landscape. **Cadastral maps** show how land is divided into real property, and sometimes the kinds of built improvements on that land. How each type of map is made with geographic information is a question that you will be able to answer generally at the end of Part 1. You can find out about the specific concepts and skills in Parts 2 and 3. The types shown here are an arbitrary selection intended to show how types of maps vary at different scales.

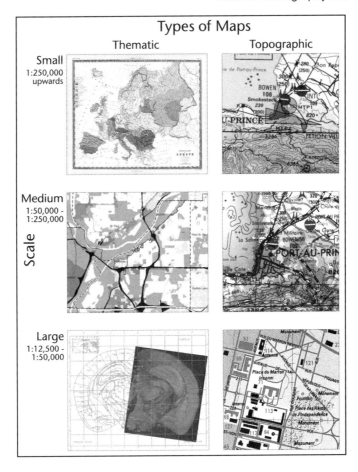

FIGURE 1.7. Maps are generally distinguished by scale and whether they are thematic or topographic in nature. Each type shown here is characterized by different geographic and cartographic representation choices.

Mental Maps

Many people find that mental maps are a great way to start thinking about how maps represent and communicate about the world (see Figure 1.8). Thematic, topographic, and cadastral maps are useful for communication because they follow known and accepted conventions, but they often have little in common with our day-to-day experiences. Mental maps are much stronger on this point, but suffer from weaknesses as a reliably understood means of communication. **Mental maps** communicate what an individual knows and can draw about some aspect and part of the world. A mental map represents particular geographic relationships based on the experience of an individual. A mental map communicates those relationships from the perceptions of one or sometimes a small group of people, but often can be

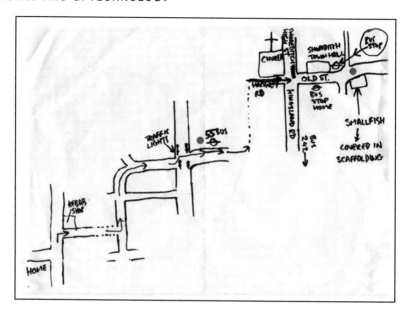

FIGURE 1.8. Mental map showing the route from home to the center of town. Note how the cartographic representation collapses space to remove or highlight significant geographic detail for this person.

difficult to understand without some form of description or use of standardized cartographic representations.

Based on human perception and behavior, Kevin Lynch developed mental maps in the 1950s as a planning technique for understanding how a city was legible. *Legible,* for Lynch, meant how well the structure and organization of a city helps support people's lives by being easily understood and requiring a minimum of effort. Using systematized graphic elements, Lynch cartographically represented people's mental maps of the city to show how they perceived and moved about the city. Mental maps are often used to help planners gain a better understanding of what features in the city need improvement or change. Many researchers have gone on to use mental maps along these lines to assess gender, race, or age differences in urban experiences and life. It is important to remember that mental maps generally lack a consistent scale or set of symbols. Because they are usually purpose-oriented and based on the selective memory and knowledge of one person or group, they are incomplete by nature and often hard for others to use. For example, in Figure 1.8, the dashed lines connecting the person's home neighborhood to downtown could indicate any distance; the readers of the map can only know how great or small a distance if they know the drawer or the area.

Geography and Cartography in Harmony

To successfully use GIS and make informative maps, geographic representation and cartographic communication must work together. Before getting into the details later in the book, let's look at how geographers and cartographers usually understand and represent the world. You may already know how your field or profession makes GI and maps. However, your work with maps and GI may greatly benefit from thinking about the conventions in your field or profession and the assumptions that go along with them. Much of this discussion can easily become part of complex philosophical discussions about existence, knowledge, and representation, but we will skip that for now to sketch out more pragmatically what many geographers and cartographers think about when figuring out how to understand, analyze, and represent the world around us. You need to get a basic idea of how maps and GI require a multifaceted framework involving complex trade-offs and balances and following many conventions. Understanding principles and conventions of geographic and cartographic representation will be the basis for considering concepts and skills required for making and using maps and GI (see Figure 1.9).

Things and Events

As indicated earlier, if something can be represented geographically, it is either a thing or an event. These are the terms most people also commonly call the elements represented on maps. You'll see later some other terms, such as feature and object, that will help you think about the possibilities

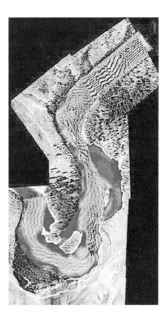

FIGURE 1.9. Image showing the results of stream modeling combining a geographic representation of events and their translation into the cartographic representation of things.

and limits of maps and geographic information. Right now, however, let's consider how things and events can be represented in GI or maps. **Purpose** is an important factor for guiding the choices made when making and using maps and GI. For an example, assume you want to make a map of places where you live where traffic jams occur. You need to show the location of the traffic jams and the roads they occur on. The map should also show the location of attractions and important landmarks to help people navigate who are unfamiliar with the area and its roads. Showing the things that are traffic jams is helpful for people visiting town, but it may not be enough if you want to help someone coming into town at 5:00 P.M. for the first time find the best way to avoid known traffic jams to the best restaurant on Main Street downtown. For this map, you need to show not only *where* traffic jams occur but *when* they occur so that your visitor knows which roads to take and which roads to avoid (see Plate 1). It would be better for the purpose of guiding your visitor to just show the traffic-jam events that can hinder his or her trip to the restaurant. Showing a lot of information about attractions and landmarks would not add to your visitor's ability to find his or her way.

Remember that GI or map patterns are static even when they refer to a process such as too many vehicles traveling at the same time. Another example would be that of a lake, which is usually a constantly present water body, but may change in size or even be completely episodic, only appearing after extremely heavy rains in an already wet season.

Animation techniques and dynamic GIS offer some interesting and advanced possibilities to create dynamic cartographic visualizations that show temporal changes. GI analysis and geostatistics provide a number of techniques for reliably representing processes (see Chapters 15 and 16).

Abstraction and Reliability

Part of what makes representation difficult is that it simultaneously abstracts while it attempts to assure reliability. **Abstraction** reduces complexity, or simplifies, to highlight essential things, events, and relationships. **Reliability** is the characteristic of a representation that refers to its dependability.

Representing things and events is complex because of the very nature of abstraction. The world is theoretically infinitely complex: it is not possible for anyone to make a map of everything in the world; nor is it possible to make a map of any area showing everything at that place, even at just one time. Every map is an abstraction that focuses on a selection of things and events from that place according to the purpose of the map. A highway map emphasizes roads and represents buildings, rivers, and towers in the landscape as context for using the map to navigate. A map of forest fires shows the location of fires and represents the slope and exposure of the hills or mountains with contour lines.

Geographers and cartographers want these representations to be reliable for the purpose they are intended for. A highway map is good for driving in a car; it is less useful for riding a bicycle and of little use for planning a trail hike in a state park. The maps of forest fires in a national park are good for understanding where forest fires occurred, but may be less helpful to determine why the fires occur and where they may occur in the future, and not of much use at all when planning a hike.

The reliability of a map or GI depends greatly on the choices made in abstracting things and events to the static patterns shown on a map or the information stored on a computer. Following Nick Chrisman, these choices are part of a framework encompassing measurements that record aspects of geographic things and events; representation of these measurements as GI to indicate geographic things, events, and associations; operations on these measurements to produce more measurements; and transformations of the representations to other frameworks. The integrity of the process of representation makes for reliable maps and GI. (You will see how this merges together into what can be called GI representation later in Chapters 3 and 11.)

Space, Things, Events, and Associations

All geographic measurements start out with observations of things and events. Things and events may seem at times to be independent of each other in a map, but geographic relationships bind them together in associations. How you approach them is not only a matter of geography or cartography, but also a matter of a field's or a discipline's conventions. A geomorphologist thinks quite differently about stream beds than a limnologist. A city planner thinks that street centerlines are good for zoning boundaries; a geodesist may differ. You may have your own examples from your field. Many GIS scientists think about these associations in terms of a predicate calculus that can be manipulated to model the situation and develop stable descriptions. Most people using GIS are glad when they get the results of this science, but they pragmatically focus on working with what they know and improving that knowledge and their abilities. This latter point is the focus of this book, although finding out about the underlying science is important to learn about too.

Geographic information science (GISci), the field concerned with the underlying theories and concepts of geographic information, is pertinent when you learn about geographic and cartographic concepts for maps and GI. You might already be familiar with the terms **spatial** and **geospatial**, which refer to properties that take place in space, especially activities on the earth. These terms refer to understandings of the world that are slightly different from geography's interest in places and spaces. The terms may suggest that the work they describe is done for purposes outside of a traditional understanding of geography. In any case, these terms, along with *geographic,* are for this book's purposes synonymous. However, you should be aware that the terms used in this book can vary in meaning among disciplines and settings.

The underlying disciplinary concepts of space, relationships, and associations also can vary greatly. Space is a continuous area. Things and events in space can be related or associated. **Related** means that the things and events are connected in terms of distance. **Associated** means that things and events occur together, without any intervening distance. This book adopts a pragmatic perspective regarding disciplinary concepts, which is related to an empirical and contextual understanding that things and events mean what they do because of who is creating the meaning and in which context.

Frameworks and Conventions

Many people look at maps dubiously. They might say maps don't make any sense; they don't match what they see; they are far too complex. Many other people, however, almost feel lost going anywhere without a map. Why is that? There are certainly many personal and subjective reasons involved, but I want to suggest that a great deal of the troubles using maps can be helped by getting to better know the frameworks and conventions of cartography established over the last 500 years by Western civilization (and before then by other civilizations). These frameworks and conventions are crucial to understanding what maps and geographic information show us, how we understand them, and how you make maps and geographic information (see Figure 1.10). You can think of **frameworks** as sets of normative and acceptable ways for showing things, events, space, relationships, and associations. Conventions are commonly accepted ways of representing and communicating in maps and GI. You can think of them as how people usually say they "do" things. Overall, cartographic frameworks and conventions vary as greatly as any fashion, but are also very distinct to a place and period.

The elements of our cartographic framework and conventions are very profound and important in ensuring that maps and GI make sense to people. Obvious examples are the almost universal north orientation of maps

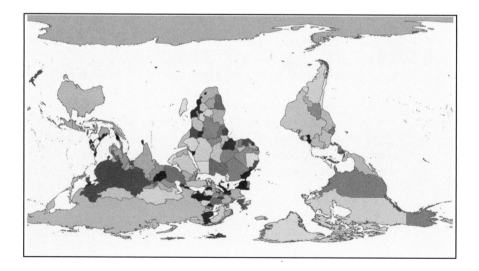

FIGURE 1.10. Even simply flipping a map of the world so that Antarctica and Australia are at the top of the map can prove confusing because of the different cartographic representation, which is very unconventional.

and the depiction of water with a shade of blue. It's not too hard to think of other examples, for example, U.S. highway symbols, how utility engineers indicate transformers and other infrastructure for line crews, how railway engineers indicate switches. But if you can look at maps from different countries, you'll start to recognize certain dissimilarities in the map symbols of each country. Even countries similar in cultural backgrounds can use symbols very differently. States on U.S. highway maps are usually drawn in a pastel color, but European maps of all of Europe choose more vibrant colors. U.S. maps generally don't show individual mountain ranges, but European maps generally show mountains and large hills with shading.

Frameworks and conventions can also serve particular professions and ideologies. Maps for specific uses, professions, and disciplines usually follow a number of conventions for simplifying complex concepts that make it very hard for untrained persons to make good maps. If you aren't familiar with the frameworks and conventions, then maps are very complex and hard to read. That's why in places where almost all people need to use a map (at some point), that map is usually very simple and shows a specially highlighted symbol where you (and the map) are. GI and most maps lack these. They are very frustrating to use without understanding the framework and conventions of maps and GI. Of course, we need also to consider the cultural biases and ideologies of maps. Many a European map of newly discovered areas conveniently erased most traces of aboriginal inhabitation; later, many Western maps showed religiously significant sites with blatant disregard. Because maps and GI are often the only sources of detailed GI, many people accept them as the best indications, even though biases may be great and the cartographic communication only partially works (see Figure 1.11).

FIGURE 1.11. A combination of maps and signs located on city streets aids people with finding their way, in many cities helping people understand the cartographic representation of the city and relate it to their own experiences. Mental maps usually reflect these understandings.

Quality and Choices

A commonly used word to describe the reliability and integrity of maps and GI is **quality**. Most people consider a highway map to have good quality if they can use it to find their way driving easily. While a highway map of the western United States may not be good to find the way to a hotel in the center of Los Angeles, this particular highway map still is very useful for finding the highways from Los Angeles to Portland, Oregon. Simply said, good-quality GI or maps are useful for the purpose we create or intend to use them for. Quality often means reliability and helpfulness for a task, but in regard to maps it often also means that the map maintains integrity regarding the world and fits the use we intend the map for. We could turn to some highway maps to figure out the size of towns in a state, but most highway maps will not help determine the extent of the towns. The concept **fitness for use** helps us get a grip on the slippery concept of quality. What this means is that we need to know the intended purposes of a map before we decide what degree of quality the map has for certain uses. You can read more about fitness for use in Chapter 12.

For now let's let quality mean fitness for use, which is the ability of the final product to support a defined task. Because of the endless number of choices people face in cartography and geography, the quality of a map or GI amalgamates the purposes, data sources, and data processing. A highway map follows geographic and cartographic representation choices that enhance its suitability for driving. While it is obvious that a map showing all of the United States, Europe, or India is not well suited for finding the

---

IN DEPTH **Data and Information: Are They Grammatical Plurals or Uncountable Nouns?**

Some of the trickiest issues in English are changes in usage that are matters of style and subject to dispute. Usually editors decide on these matters, following conventions or stylistic preferences, and writers follow them, or they are brought into line by copy editors. In the case of data and information, two central terms for this book, I find it important to note, following the discussion of language changes in Garner's *Modern American Usage* (2009), that data increasingly is treated as a mass noun taking a singular verb; however, in some formal contexts data still is preferably treated as a plural. In computing and allied fields the singular use of data is widely accepted. *Datum*, the grammatical singular form of the plural *data*, has a special use in geography and cartography and in fields where it also refers to a measured or calculated reference value and location. It is used little otherwise. In any case, Garner predicts that in roughly 50 years we will all use data in the singular. Information is more straightforward and already is used widely, if not always, as a mass (or noncount) noun. It cannot be enumerated. Although we cannot be certain if this book will be read in 50 years, I will use both data and information as mass nouns in the book with an eye toward the future.

nearest restaurant from an office, house, or hotel, many maps and much GI are intended for multiple uses, which will give them both strengths and weaknesses. GI may be collected and prepared for purposes that invite many different uses, but it may in fact be ill-suited for particular purposes. In the end, most maps are quite limited: a highway map is not very helpful for hikers (except for helping them get to where they want to hike). Due to its malleability GI can be handled more flexibly, but this does not remove significant limits. Well-known errors have occurred when people creating road databases included ferry routes as connections for national roads and then entered them as roads in a car navigation system. More than one driver has taken a surprise bath as a result.

Some of the many choices for map and GI quality are fundamental, which means they occur regularly. All are topics that this book examines in depth. As you get a better grip on these choices, you will better understand the quality of maps and GI and also how to create better maps and geographic information. Below is a list of important topics and the chapters in which they are covered.

> *Projections:* How geographic locations on the round earth are shown on a flat map or coordinate system is known as **projection**, and it is one of the most important choices that affect quality (Chapter 5).
>
> *Coordinate systems:* Related to a projection, a coordinate system is especially pertinent for GI, which can easily be combined with other GI that uses the same projection and coordinate system (Chapter 6).
>
> *Symbols:* Pictures or other graphic patterns that stand for things map users can recognize are **symbols**. For most people using maps, the symbols are the most important choice, because it may affect whether they can make sense of the map (Chapter 11).
>
> *Geographic representation:* Deciding how to show things and events is crucial to whether a road can be modeled with different lanes of traffic and sidewalks or only in terms of traffic flow (Chapter 2).
>
> *Cartographic representation:* If the geographic representation provides the information, the cartographic representation can support various representations, contingent on a number of parameters, notably scale (Chapter 2).

Conventions and Quality: An Example

You can probably think of a number of times when a map you had wasn't as helpful as you wished it would be. That may well have been because of quality and choice issues. Before moving to more specifics concerning communication and representation in the next chapters, the following example shows how frameworks and conventions in relation to quality and choices can lead to a less-than-useful map. This is a very big problem for companies whose business depends on maps, so they put great effort into making maps understandable, but have to make some important choices that may greatly limit the quality of their maps for some groups of people.

People who travel often have to use rental cars and depend on their maps (see Figure 1.12). If you arrive in a country with a strong mapping tradition and rent a car at the city airport, you may receive a very good road map of all of the country; this map may be perfect for finding the smallest town that your friend's ancestors originated from, but is probably much less perfect for finding the way to your hotel. But let's assume that you have been given a pretty good map of the city, one that even shows hotels. It's no problem to find the hotel you're going to . . . but wait. The center of the city is on the map, the hotel is there, but where's the airport? What do you do now? Where is the airport? What's the road to the city called? The framework for such maps is suited for people who know the city and speak some local language and who are looking to find their way around, not for visiting businesspeople or tourists. In this example, not at all uncommon, you can summarize that the conventions for the local maps support the general orientation of people who are familiar with the conventions of these maps and the culture, not the specific conventions and definitely not the purpose of a visitor from another country.

Distinguishing GI from Maps

Maps remain important, but more and more maps are produced with GI and used online on computers, tablets, or mobile phones. Some people now even suggest that most maps are simply interfaces to GI databases. Several

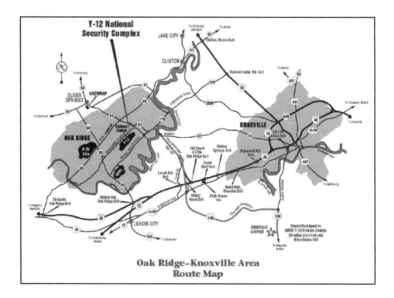

FIGURE 1.12. A directional map can be useful for finding some directions, but its geographic representation will fall short in supporting other uses, such as finding the way from the airport to a hotel.

years ago, separating GI from maps would have been complicated. Maps, following the International Cartographic Association, are science and art. GI is interpreted or symbolized data. It's simpler now. In this book **maps** are a form of output of GI. Maps are truly the most common form of output and have been essential to our understanding of the world for millennia. Maps can be drawn by hand, and constructed by hand, but nowadays are mostly prepared using GI (see Figure 1.13).

The computerization of cartography changes the possibilities you have for working with the underlying GI. GI is usually presented as maps, but tables, figures, and hybrid output forms are also legitimate. GI is what is used in GIS. Data is what information is before it is used and makes sense to the persons creating or using the GI or map. Earlier in this chapter, I compared GI to oxygen. Now, starting there, you can think of one of the effects of information: information has at least the potential of having an effect. Data may sit in an archive for years and years—never having an effect until someone looks through the data, makes sense out of it, and converts it to information.

GI is not data (see Figure 1.14). Though data can become information, or at one time it may have been information, correctly understood it is only the raw recording of measurements. Such data then are used to create information. To become information, measurements must be put into relationships with each other and with a purpose. The way the same measurement, for example, a country's border, can be used in two different projections (see

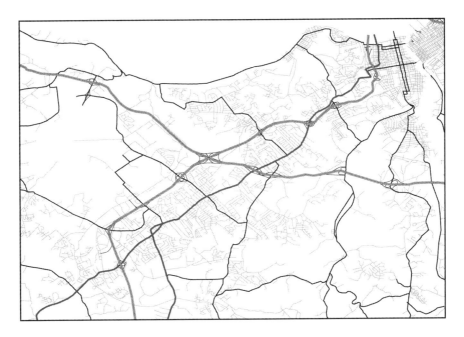

FIGURE 1.13. An example of vector line data using a cartographic representation that distinguishes road types.

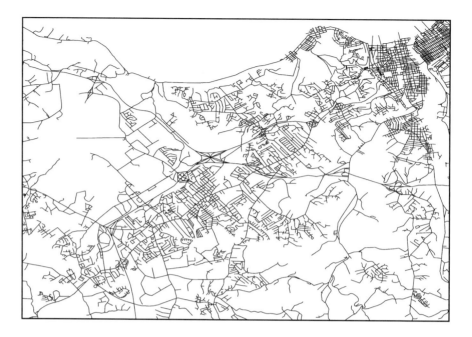

FIGURE 1.14. The same features from Figure 1.13 shown without any cartographic representation.

Figure 1.15) illustrates this difference. For one person, data used unchanged may be merely data, but another person may use that same data as the basis for creating information. How do we know which is which? Apply the sense test. If what you see, regardless of its form, makes sense, it is information because sense, or meaning, only comes if the thing you see has an effect; if it doesn't make sense, it is just data. Of course, this only means it makes sense for you. Making the same sense for others is a much harder, but more important, test that we should use to assure the maps we make communicate what we want.

Summary

GI and maps are ways of representing what people see and observe. Things and events should be distinguished. *Things* refer to static representations of something in the world; *events* refer to dynamic changes. Because of the complexity of the world, even small aspects of it, and the challenges of representing the world, a number of choices are brought together in geographic representations and cartographic representations. Because of the large number of choices, cultures, fields of science, and professions rely on conventions and frameworks. Conventions are often unstated guidelines for representation. Frameworks are rules and procedures for dealing with the

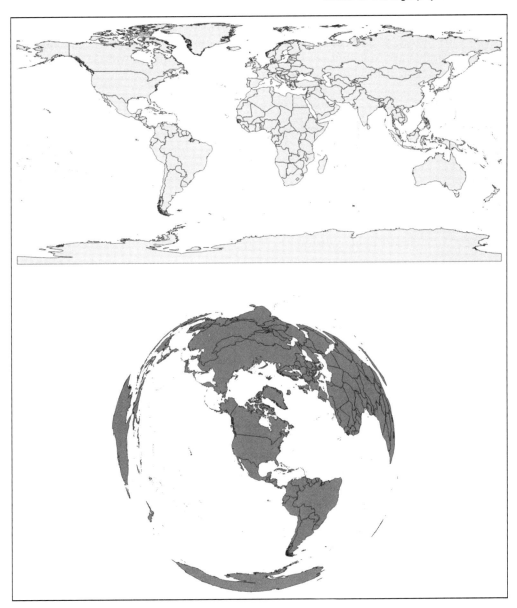

FIGURE 1.15. GI showing boundaries of countries can be projected in many ways. Both the top and bottom panel of this figure use the same boundary information, but create different projections.

complexity of choices. Geographic information and maps have a great deal of power as a result of making portions of the world understandable. The quality and reliability of GI and maps depends on how well they fit the purpose, or fitness for use.

REVIEW QUESTIONS

1. What distinguishes a *thing* from an *event*?

2. What is the geographic significance of the difference between a *thing* and an *event*?

3. How are things represented in cartography?

4. What influences the quality of GI or a map?

5. Is every map reliable enough for every use?

6. How do representation and communication relate to each other?

7. Why can't GI or a map show everything in any chosen area without distortion?

8. What role do conventions play in the creation of GI and maps?

9. Can GI or a map be understood just by itself?

10. What are the differences between GI and maps?

11. What makes maps so powerful?

ANSWERS

1. What distinguishes a *thing* from an *event*?

   A *thing* is a geographic representation that is static and nondynamic; an *event* is a process or part of a process characterized by a change in how it is being described and accounted for. This difference is significant for cartographic representation and cartographic communication. Things can be easily portrayed with cartographic symbols on paper or other media; events can be only portrayed as a series of "snapshots" or through animations.

2. What is the geographic significance of the difference between a *thing* and an *event*?

   The difference between a *thing* and an *event* is significant geographically and also for cartographic representation and cartographic communication. The geographic representation of things emphasizes consistent characteristics; events emphasize changes or processes.

3. How are things represented in cartography?

   Things can be easily portrayed with cartographic symbols on paper or other media; events can be only portrayed as a series of "snapshots" or through animations.

4. What influences the quality of GI or a map?

   The choices made that affect the reliability and integrity of the GI or map.

5. Is every map reliable enough for every use?

   No, maps are abstractions that focus on a selection of things and events from the endless complexity of the world.

6. How do representation and communication relate to each other?

   Representation is the basis for communication. It is impossible to have a cartographic representation of what is missing from a geographic representation.

7. Why can't GI or a map show everything in any chosen area without distortion?

   Maps must abstract, which leads to distortions. Some distortions are explicit, but many can be implicit.

8. What role do conventions play in the creation of GI and maps?

   Conventions are simply unstated rules and assumptions that people rely on to help with the geographic and cartographic representation of the world. They are often a big part of what people "do" to make a map.

9. Can GI or a map be understood just by itself?

   It may seem this way, but that understanding arises because of the knowledge we have of conventions—for example, water is blue, north is usually at the top edge of the map, and so on.

10. What are the differences between GI and maps?

    Maps are printed or displayed on a media that cannot be changed or altered without altering the map. Maps cannot be altered except by destroying the map and reusing portions of it for other purposes. GI can be used over countless times, in different ways, to make different maps.

11. What makes maps so powerful?

    Much that we know is known to us only through maps. The world, Asia, the United States, even an entire city are places of which we can only experience a fraction of. Maps put things and events in the world into a comprehensible graphic format that communicates. Maps are powerful when they successfully communicate what we didn't know before.

Environmental Monitoring in Central and Eastern Europe

Dr. Marek Baranowski has served as the director of the Warsaw, Poland, office of the United Nations Environmental Program, Global Resource Information Database (UNEP-GRID), since 1991. He is a geographer by background and has worked with GIS since 1973. This office is involved in many environmental projects in Central and Eastern Europe, which since 1989 have seen rapid change. Educational and learning resources prepared by UNEP/GRID-Warsaw have twice won (in 2001 and 2003) first-place awards from the Polish Ministry of Environment for outstanding achievements in environmental science and development. He and the office are also involved in several national and regional environmental monitoring projects, visualization for participatory planning, the Polish general geographic database, and EuroGlobalMap.

GI and cartography are central to these projects. The educational and learning resources use GIS to collect and prepare data; maps are central to the multimedia educational tools used in classrooms across Poland. The coordination of national and regional environmental monitoring involves GI and maps. Working on the Carpathian Environmental Outlook, which is connected to the Global Environmental Outlook, requires the coordination of information from the countries of Austria, Czechia, Hungary, Poland, Rumania, Serbia-Montenegro, Slovakia, and Ukraine. Because they need to show past changes, they must pay careful attention to the seasons, time, types, and resolution of remote sensing data used to detect changes. When we spoke, Dr. Baranowski told me that a previous project of the CORINE Land Cover project in Romania once had to be repeated in order to recollect all the data on the basis of new satellite images since so many changes had taken place between the start and the conclusion of work.

GI and cartography have limits for showing dynamic processes. According to Dr. Baranowski, detection of changes is the key way to show processes in this large area. Each human–environment interaction can not be detected and recorded in an area where over 15 million people live. Detecting changes becomes complicated because of different data collection issues. For example, satellite images collected in spring are different than images collected in the fall. Large errors in the determination of changes could result when using noncomparable images. Detailed technical specifications address these issues and other concerns. The challenges of dealing with human–environment interactions at this scale involve technology as much as organizations.

More information about UNEP/GRID-Warsaw is available at *www.gridw.pl*.

Chapter Readings

Board, C. (1967). Maps as models. In R. Charley & P. Haggett (Eds.), *Models in geography* (pp. 671–726). London: Methuen.

Buttenfield, B. P. (1997). Talking in the tree house: Communication and representation in cartography. *Cartographic Perspectives, 27,* 20–23.

Chrisman, N. R. (1997). *Exploring geographic information systems.* New York: Wiley.

Dorling, D., & Fairbairn, D. (1997). *Mapping: Ways of representing the world.* Edinburgh Gate, Harlow, UK: Addison Wesley Longman.

Gersmehl, P. J. (1985). The data, the reader, and the innocent bystander—A parable for map users. *Professional Geographer, 37*(3), 329–334.

Kaiser, W. L., & Wood, D. (2001). *Seeing through maps: The power of images to shape our world view.* Amherst, MA: ODT.

Carl Steinitz and Landscape Architecture

Landscape architects deserve special mention in the history of GIS. They were involved in developing many of GIS's first key applications including techniques based on traditional overlays of transparent thematic maps on a topographic base map. This overlay technique allowed the landscape architects to provide a situational map that could be shown with specific themes (parks, schools, ecotones, etc.) and with composites of the themes. This became a key part of GIS-based analysis. Ian McHarg was one of the key developers and promoters of the overlay-based technique for planning. Carl Steinitz, a contemporary of McHarg, stands out for his contributions to the development of overlay techniques, their application, and documenting the history of overlay techniques in landscape architecture.

Further Reading

Steinitz, C., Parker, P., & Jordan, L. (1986). Hand-drawn overlays: Their history and prospective uses. *Landscape Architecture, 66*(5), 444–455.

Monmonier, M. (1993). *Mapping it out: Expository cartography for the humanities and social sciences.* Chicago: University of Chicago Press.

Monmonier, M. (1995). *Drawing the line: Tales of maps and cartocontroversy.* New York: Holt.

Pickles, J. (2004). *A history of spaces: Cartographic reason, mapping, and the geo-coded world.* New York: Routledge.

Thompson, M. (1987). *Maps for America.* Washington, DC: U.S. Department of the Interior, Geological Survey.

Wood, D. (1993). The power of maps. *Scientific American, 268*(95), 88–93.

Web Resources

🕐 National Geographic's online Education offers a highly interactive introduction to geography and cartography. The Mental Mapper offers a good introduction to mental mapping and you can follow one of the lesson plans. See *http://education.nationalgeographic.com.*

🕐 Three hundred and twenty-one definitions of the word *map* are collected at J. H. Andrews's website and analyzed visually by John Krygier: *http://makingmaps.net/2008/11/25/321-definitions-of-map.*

🕐 The ICA website provides up-to-date information about cartographic research, teaching, and publishing around the world: *www.icaci.org.*

🕐 The University Consortium for Geographic Information Science (UCGIS) website offers information about some of the newest GI research: *www.ucgis.org.*

🕐 For many, the USGS website is the first place to go to for information about mapping in the United States: *www.usgs.gov.*

EXERCISES

1. Comparing Map Representations and Our Observations

Get a map for a well-known area. The map can be at any scale or of any type; it should show things that people are well familiar with. Start out by comparing what you remember of that area to how it is represented on the map. What is missing? What is simplified? What has been added or exaggerated? Compare your map and answers to these questions to a neighbor's. Are they the same or different lists? How could that depend on scale or the type of map?

2. Make Mental Maps and Discuss Them with a Classmate

Draw maps from memory of an area you and your neighbor are familiar with. When you are done, identify common elements with your neighbor or in small groups and write them down on the board in your classroom and discuss how well the mental maps help communication. Also discuss how the maps could be drawn differently and what cartographic or geographic choices the different maps involve.

3. Choices and Scales

Examine the figure showing the three types of maps and different scales. What choices do you think were made to make each map? What is the scale of each map? How do the differences between map scales affect the way things are shown? How do the choices differ in relation to scale and map type?

EXTENDED EXERCISE

4. Mental Maps

Objectives:

> Communication with maps
> Things and events as patterns and processes
> Accuracy is related to use

Overview

Mental maps are a way of portraying geographic relationships and features we remember cognitively but in practice more often should be better called sketch maps. In this exercise you will prepare a mental map of the area you live in. Later, you will find a map of the same area on the Internet and compare the two maps in terms of what they communicate and their suitability for navigation.

Instructions

On a piece of plain white paper draw your mental map of the area from memory. Don't just draw (or copy from) a street map! This should be a mental map, not a cartographic map. Take about 20 minutes to draw and annotate (for your instructor's sake) your map. Show as much detail as you can, and use a different color for the annotations if you can. Remember to focus on making the map accurate only in terms of what is important to you—the places where you live, eat, work, walk, recreate, and so on. Leave off things that are not important. Include the following elements and symbols:

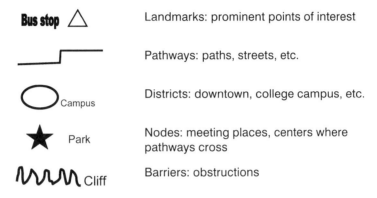

Bus stop △	Landmarks: prominent points of interest
‾‾‾⌐	Pathways: paths, streets, etc.
◯ Campus	Districts: downtown, college campus, etc.
★ Park	Nodes: meeting places, centers where pathways cross
∿∿∿ Cliff	Barriers: obstructions

Communicating with Maps

Find a road map of the same area on the Internet. You can find maps for most areas at several websites including *mapquest.com, maps.yahoo.com*, and *www.multimap.com*. Print the map out if you can. What are the key differences between your map and the online map? Are both maps showing the same things? How can you explain the differences? Does it have something to do with the reason for making each map? Which map is better for communicating?

Patterns and Processes

One of the most interesting things about maps is that because the paper and drawing won't change by itself after you make it, you have to show things, such as a house or store, in the same way as you show processes—for example, the way you walk to a bus stop or drive a car to work. If you use multiple colors you can separate things and events, but remember the map doesn't show the process, it only shows an event that corresponds to the process.

Accuracy Is Also a Question of Use

Comparing the two maps, it seems to make sense that neither map is better than the other for communicating. If you want to explain to someone the place you live in, your mental map is much better in communicating the places you like, where you live and work, and what is significant in this area for you. If you just needed to explain to a visitor how to get to campus, downtown, a store, or park, than the online map is probably better suited because it focuses on giving the information needed for navigation in the area.

The potential use of a map is an important factor in determining the map's accuracy. While the online map is more accurate for general navigation, your mental map may be better for explaining to a visiting relative how to meet you at the local park or café you frequently go to.

Questions

1. What do you personally consider to be the most important features you drew on your map? Why are they important?
2. Are there blank areas on your map? If so, why? What do you guess is in these "empty" spaces?
3. How long have you lived in the area? How has this affected your mental map?
4. Do you use a car? A bicycle? How does this affect your mental map?
5. How does your mental map compare to the road map? Consider differences in detail and the use of the maps for navigation. What purposes do you think each map is better suited for?

CHAPTER 2

Choices in How
We Make Representations

As you read in Chapter 1, geographic representations and cartographic representations are abstractions. Any representation of the world is always an abstraction. It reduces complexity, simplifies, and highlights essential things, events, and relationships. An artist's painting, Hollywood movies, and a child's drawing of home are all obviously abstractions. GI and maps are also abstract representations, but they may be less obviously so, for a multitude of reasons. Most people believe maps are more accurate than paintings, for instance, and they wouldn't question a map's representation in the way they would question a painting's representation. One reason for this is that maps generally follow unstated rules and notions, or conventions. The established conventions for GI and maps explicitly and implicitly guide people in making choices and reinforcing ideas that GI and maps are more accurate. GI and maps follow many conventions to ease understanding in order to facilitate communication. In other words, abstraction isn't "bad"—it's necessary for the sharing of knowledge and information. However, even with conventions that implicitly influence many aspects of cartography and GI, many map and GI abstractions remain difficult to understand until we have specific contextual information—for example, an electrical utility's power-line map may come with explanations, but most of us would not understand either the map or the explanations.

It is certainly true that a great deal of the ease of "reading" maps comes from an individual's familiarity with conventions. Once you have understood the conventions, understanding maps is much easier and it even becomes hard to imagine how one could go without the insight they offer. An infinite number of choices are possible when creating GI and maps, so conventions play a key role in limiting choices to make understanding easier. You can find out about the specifics of GI representation types in Chapter 10. How well geographic information and maps communicate greatly depends on how they apply established conventions. The world-at-night image in Figure 2.1 would be much harder to understand if the outlines of the continents were less clear: it would be difficult to see which parts of a continental land-mass aren't strongly lit if the continents themselves were not shown in a shade of gray. Understanding the principles that are implicit in conventions can also help you understand maps better and prepare you for working with GI and maps.

This chapter familiarizes you with the underlying concepts of geo-graphic and cartographic representation and cartographic communication. For now, you should keep the two types of representation separate from each other in the following sense: *geographic representation* deals with how people choose aspects of the world to show on a map or as GI; *cartographic representation* is the process and choices involved in going from a geographic representation to the symbols to communicate with readers. You already should be aware that as we become more versed in the issues, they often melt into one. Many people refer to them together as **modeling**, other people will distinguish information modeling from cartographic modeling. As we learn the specifics of GI and cartography, holding them apart will help you to learn the underlying concepts and skills required, the role of conventions, and the various meanings of models. These concepts and skills are, of course, fundamental to understanding and applying the conventions used for GI and maps. In the following chapters you can explore the various components of

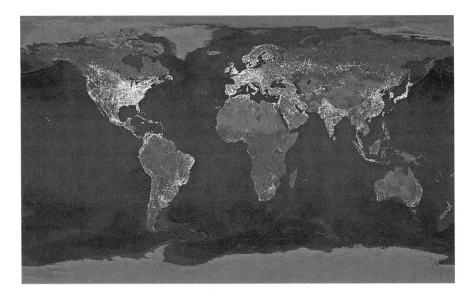

FIGURE 2.1. The world-at-night image is an abstract cartographic representation. At any one time it is physically impossible to see the entire world at night. Second, the location of many lighted areas is interpolated. This unusual combination but implicit use of cartographic conventions makes for an intriguing map.

geographic and cartographic representation in relationship to communication in greater detail.

Geographic Representation

To illustrate the issues and concepts related to geographic representation, we will begin with an example that you may already have encountered yourself: flooding. You will come back to this example throughout this chapter, so you need some background information to aid you in making connections to cartographic representation and communication issues later. After presenting some background information, we will turn to the issues in creating a geographic representation using a database. As you will see, geographic representation is fundamental to GIS, but even without a GIS you can engage in geographic representation.

The example to consider is about one river out of countless rivers that frequently flood, in this case flooding an urban area located in northern Illinois near Chicago where I grew up. The Des Plaines River is a muddy and slow river draining areas that were once prairie and now are largely developed. It flows into the Chicago River and then into the Mississippi. Children are afraid of the river because of legends about leeches and snapping turtles supposedly living there—never mind the pollution. Most adults are more

---

IN DEPTH The Parts of Maps

If some people had their way, every map would always include five elements that aid in understanding by whom, why, and when a map was made. However, like all recommendations these five parts are suggestions, not requirements. The five essential elements of maps are:

Legend (special note re color)
Scale
Orientation
Neatline
Title

Additional and important elements include:

Name of author
Date map published
Explanation of purpose
Projection
Data sources
Gridlines

afraid of the river because the river floods frequently in the spring following winter snow melt and the first rains of spring. Experts say this is because of the considerable development near the river in northern Illinois, which has left areas that were previously fields and forests covered with impervious surfaces including asphalt, concrete, and houses. Previously, open ground had absorbed most of the rain and melting snow, only slowly releasing it into the nearby Des Plaines River. Now much of the melting snow and rain flows through culverts and pipes almost directly into the river, vastly increasing the flow of water that very quickly enters the river, increasing the volume of water in the river during a storm or longer rainfall, and causing the river to overflow its banks and flood lower lying areas, even the areas that previously hadn't seen flooding for many years.

The primary purpose for the geographic presentation of flooding is to conceptualize the necessary data to help answer the questions of when and where the river can flood. A very simple model may only consider observations of river bank elevation and water height. With observations of these characteristics, we can create GI that is the basis for analyzing where the river floods. To make sure that the observations can be related to one another, special attention to the measurements must be given. First, we need to consider the relationship between water height and river bank elevation. For the entire length of the river, we need to have a defined value that indicates the level of water that leads to a flood. This value isn't simply zero because the river banks become lower as the river flows away from its higher source to its mouth on the Chicago River. The water height and river banks need to be modeled in a relationship that also remains valid if there are changes along the river bank—for example, building a dike of sandbags. An ideal choice is a system for recording elevation independent of water height and riverbank elevation. Both measurements can be related to each other and will show that if the water height is higher than the river bank elevation, a flood results; additional data can also be added later. The elevation reference needs to be explicitly defined, something a regional, state, or national mapping agency or a geodesy agency generally provides and keeps current.

And what about the question of where the river floods? The geographic representation could split the river into segments, the smaller the better, and for each segment record the average water height and river bank elevation. Using smaller segments is better because that will allow us to more accurately say where the flooding will occur, but this will require more data—which will add to the project's costs. Additionally, moving on to an issue for cartographic representation, segments that are too small will make it difficult to show where flooding occurs over the 250-mile length of the river and its tributaries on a small screen or paper because the small segments would be hard to visually distinguish. Before going on to consider more specifically how measurements and observations are geographically represented, please note that this example leaves out a number of important details—most importantly, the measurements and observations involved in determining the capacity of the river.

Measurement, Observations, and Relationships

We have at this point a simple geographic representation, or model, to indicate river flooding that considers the relationship between water height and river bank elevation. Each of the elevation measurements is related to an independent elevation reference. The river segments are related to an independent coordinate system, which can be used to assure that all observations are recorded at the same location. To place the measurements into a geographic representation used for GI we finally need to create attributes in a database that records the values for each river segment. The observations (field or calculated data) can then be stored for each segment. These values can then be compared to determine if and where the river floods for a particular water height.

IN DEPTH **When Does Data Become Information?**

This question should have a straightforward answer, but the issues at hand (representation, use, knowledge, meaning) are anything but clear. The resources identified in this chapter point to relevant literature. For simplicity's sake, you can say that *information is data that means something.* In other words, information makes sense for the uses to which it is being applied.

And data is digitally stored observations and measurements. Usually when data is collected it means something, but when data is stored in a database and eventually manipulated, transferred, copied, and even used by other people who weren't involved in collecting it, the data may have been disembodied from its meaning. To make data "mean something," it has to be associated with a context—for example, what types of measurements it is, rules for how observations were recorded, and the uses for which the data is intended. With the Internet and online data clearinghouses, inclusion of such contexts along with data have become commonplace. In spatial data infrastructures, a common source for much data online, metadata often plays an important role in facilitating the transformation of data into information.

Turning data into information is a process that goes beyond merely adding a column of like measurements. The meaning of data may be more than a set of data's collected attributes; the meaning shows itself when we consider relations the data has to a larger context (broadly defined). Again, information has meaning, but data holds the potential of gaining meaning when used in this way.

If you view a set of geographic information in a different context, most people would still call that information because of its potential to have meaning. This often gets confusing. To be specific, at least to summarize the discussion here, data can be (and often is) *called* information even though it may have reverted to the status of mere data.

The determination of flooding is based on a relationship: if the water height is greater than the river bank elevation, flooding results. This can be modeled mathematically as $w > e$. This relationship will never be stored in the geographic representation; we have to calculate it using the recorded characteristics. However, the stored geographic representations were only determined based on an understanding (however simplistic) of the relationship between water height and river bank elevation. The relationship is central for representing the flooding, but the information recorded in the geographic representation lacks this relationship at first—it must be determined and recorded as data in another step using the measurements.

In the end, measurements, observations, and relationships are all parts of geographic representations (see Figure 2.2 for more examples), but relationships are usually separate from measurements and observations and not noted together. Some relationships can be determined using measurements and observations. Remember that geographic representations represent selected aspects of things and events, which means far more than merely "storing" the data. The cartographic representations that follow provide ways to show the relationships. After all, most maps don't only show us just things or events, they show us how things affect each other or can be related.

	Description	Common Examples	Map Examples	
Nominal	qualitative measurements	name, type, state	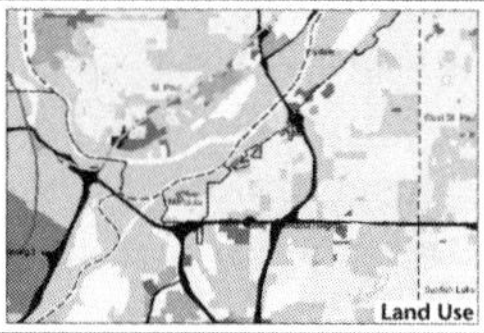 Land Use	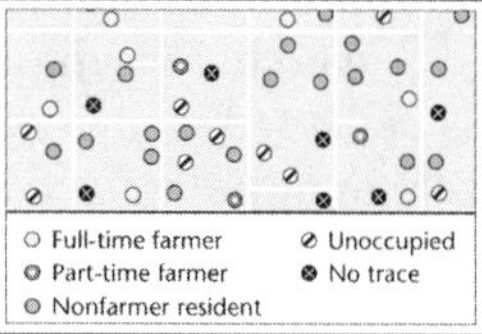 ○ Full-time farmer ⊘ Unoccupied ⊙ Part-time farmer ⊗ No trace ⊙ Nonfarmer resident
Ordinal	quantitative measurements with a clear order but without a defined zero value	small, medium, large	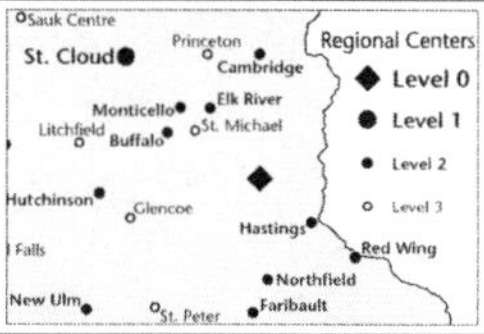	 Roads
Interval	quantitative measurements with a defined beginning point	temperature, height, distance	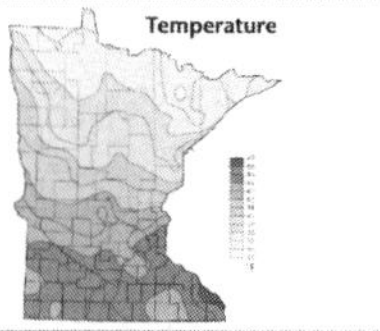 Temperature	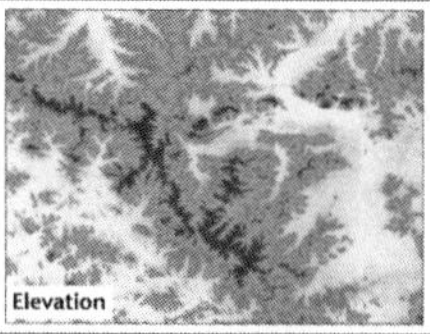 Elevation
Ratio	quantitative measurements that provide a relationship between two properties where the 0 value indicates the absence of the relationship	particulates mg/m³, time to cover a distance, dissolved oxygen in a liter of water, population density	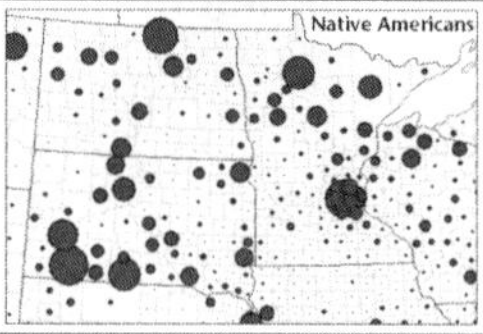 Native Americans	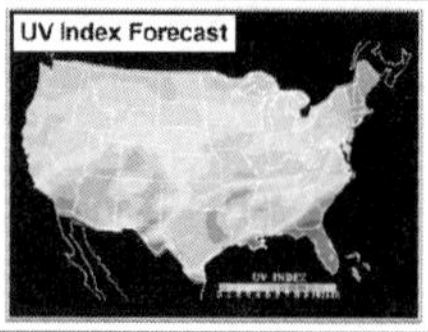 UV Index Forecast

FIGURE 2.2. Measurement types with map examples used in geographic representation.

Types of Measurement

Measurements have to be stored as values to be information. Water height is stored as an interval value because it is related to a defined starting elevation of 0, which in this case is taken from the elevation reference. The storage of measurements is a key issue for creating GI and relies on the types of measurement developed by Stanley Smith Stevens.

Stevens developed his reference scheme in the 1940s. His intent was to offer a framework for psychologists and other social scientists that could take intrinsic properties into consideration. *Extrinsic properties* are those that are directly empirically measurable: width, height, depth, elevation, and the like. *Intrinsic properties* are characteristics that can be observed, but must be associated with other properties—for example, color, age, form, and quality. Extrinsic properties can be established directly from an object, but intrinsic properties must be indirectly measured, inferred, or interpreted.

Because of the nature of intrinsic properties, Stevens proposed that measurements should be distinguished according to the ability to combine them with other measurements. For example, the measurement of a person's height cannot be meaningful combined with the measurement of his or her hair color. Following Stevens, height is an interval measure and color is a nominal value.

In all, Stevens differentiates four types of measurement, which unfortunately are not exhaustive and fail to include common types of GI such as radial measures of angles. The four measurements and their definitions are:

Nominal   Qualitative measurements (name, type, state)

Ordinal   Quantitative measurements with a clear order, but without a defined 0 value (small, medium, large)

Interval   Quantitative measurements with a defined beginning point (temperature, height, distance)

Ratio   Quantitative measurements that provide a relationship between two properties where the 0 value indicates the absence of the relationship (particulates mg/m^3, time to cover a distance, dissolved oxygen in a liter of water)

Sinton's Framework and Geographic Representation

Applying Stevens's measurement framework to manual cartography was relatively straightforward because of an individual's (or organization's) control of the design and drawing process. The use of computers and sharing of GI changed this because now data collection, management, and output are divided between many more individuals and organizations—often even without any contact to each other. During the early days of GIS, a number of people working in this area realized that the established "art and science" of the cartographer required more detailed descriptions of cartographers' and geographers' work if people were ever to successfully automate cartography, especially if maps were to be used as the basis for analysis. "Analytical cartography," as it is called, produced a number of important approaches and concepts that became critical to the success of GIS. John Sinton was an important and active contributor to this group.

Sinton devised a scheme for considering space, time, and properties in three possible roles: fixed, measured, and controlled. When space is fixed, a measuring device (e.g., tide gauge, stream gauge, NO_2 monitor, etc.) measures an attribute at a set interval of time (e.g., constantly, every 10 minutes, once a day, etc.). Measuring devices often use time intervals as the control. They produce information about a single place. People usually collect GI about multiple places by fixing the time and measuring characteristics of space or some other attribute, and controlling the space or attribute.

The distinction between what is controlled and what is measured is important for geographic representation of vector and raster data. If the attribute is fixed and the space is measured, the resulting representation is a **vector representation** showing the extent of the attribute; if the space is fixed and the attribute is measured, the resulting representation is a **raster representation**. Of course, there are exceptions and limitations to this approach, which is why we say it is the way that GI is *usually* collected. First, point data—for example, the location of the stream gauge or a measurement of soil pH, or the location of a truck—does not show extent, but it does show location based on the measurement of space for a fixed attribute. Second, this scheme gets very complex if even everyday objects with many measurements are considered. Third, and most importantly, Sinton's scheme does not take into account relationships (see Figure 2.3). As in the river flooding example earlier, implicit water levels can only be related by use of common elevation base values. Sinton's framework is very helpful in making distinctions between raster and vector geographic representations (see Figure 2.4).

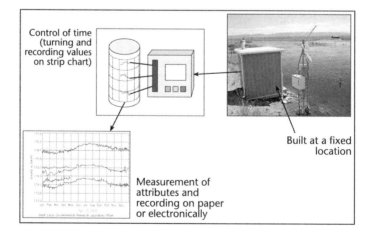

■ FIGURE 2.3. Stream gauge showing fixed location, measured attributes (strip chart), and controlled time (minutes/hours) following Sinton's framework.

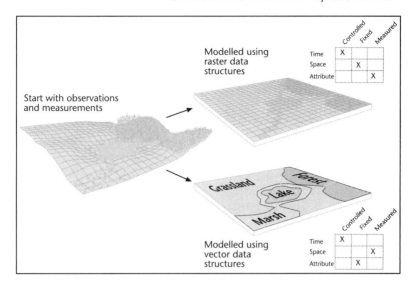

FIGURE 2.4. Raster and vector geographic representations and Sinton's framework.

Choices and Geographic Representation

The choices involved with creating geographic representation are wide ranging and often elusive because conventional ways of understanding geography and cartography lump them together. Further, some of the choices have direct and obvious consequences for cartographic representation and communication, but the consequences of others are hard to pin down. For each of the choices identified here, you will find an indication of how it is relevant (highlighted in italics) to the example of river flooding we examined earlier in the chapter. These choices are discussed in more detail in Chapter 3 and the following chapters.

Data collection   To represent a thing or event, **data collection** must take place. The collected data must offer sufficient geographic, attribute, and temporal detail for the intended purposes and uses.

How are the measurements of water height and river bank elevation made?

Data updates   Changes in land use, roads, and so on impair reliable communication with GI and maps. Basing a project on somewhat out-of-date source material may be inevitable, so a plan for incorporating **data updating** should be part of a project.

How can changes to the river banks and drainage be considered?

Attributes	What characteristics and qualities of things and events are included and how they are recorded make certain representations and analysis possible or impossible.
	Should changes to water height be recorded as new attributes or should they replace the existing attributes?
Coordinate system	Commonly used for GI, maps also make use of coordinate systems, a combination of a projection, datum, and locational reference system. The coordinate system is an especially important choice for GI and in some areas may even be legally defined.
	What is the best coordinate system for showing the river with sufficient accuracy and detail?
Vector/raster	Will the map emphasize the areal extent of particular attributes (vector) or the presence of particular attributes (raster)?
	Recorded as vector data, the areal extent of each segment is clear, but is the raster data perhaps more advantageous for observing and measuring?
Data combination	Data for a project usually comes from more than one set of data. Choices of projection, scale, and coordinate system become especially important. Having a plan for how and when to implement **data combination** is, therefore, necessary.
	It is of the greatest importance that water height and river bank elevation measurements be recorded geographically using the same coordinate system.

Cartographic Representation

In traditional cartographic mapmaking, which lacked a distinct phase of working with GI collected by others, cartographic representation was a choice of map elements, level of generalization, and visual variables, as well as many of the geographic representation elements. The measurements and observations could be transformed manually into a dizzying array of graphical elements, some of which still remain beyond the means of GIS or remain highly difficult to create even with more specialized applications. In most GIS work, some or even all of the data has already been prepared, making cartographic representation often a more distinct part of the process.

For work with both GI and maps, the basic choices remain the same, but it's fair to say that you will have fewer chances to work directly with observations and measurements. Because of the availability of GI, in most cases, people are constrained by the geographic representation. Of course, there is some flexibility, but for the most part options are decisively limited by choices already made for the geographic representation. This limited flexibility and the options that go along with it are discussed in greater detail in the following chapters. The following choices serve only to orient you and are related to the river flooding example from earlier in the chapter to help you better understand the relationship between geographic and cartographic representations.

Choices for Cartographic Representation

Projection	Every map made on a flat sheet of paper or shown on a flat screen is projected. An infinite number of projections can be made. Fortunately, in most cases at least, commonly used projections have been established.
	Which common projection used in northern Illinois is best suited for the presentation of river flooding? This issue needs to be considered together with the coordinate system.
Scale	To represent locations measured in a projection system to locations on a map or coordinates used for GI, the locations are scaled. Scale also constrains how things, events, and geographic relationships can be shown.
	Which scale is best suited for showing the river flooding? What scale best balances detail for the cartographic products required?
Points, lines, areas	Vector data can be easily transformed between these geometric types, expanding cartographic representation possibilities.
	The river, geographically represented as a line, could be converted to an area. Would this help people understand river flooding better or confuse them into thinking the river is much wider than it is?
Cells or pixels	Raster data can be shown either as cells, which take up an area, or as a pixel, which is a single dot, much like the dots making up the image on a TV screen, which combined show an area.
	If the river is geographically represented as raster data, does the presentation of the river as cells or pixels help the cartographic representation?
Symbols	The symbols for elements from the geographic representation can be varied in terms of size, shape, value, texture, orientation, and hue.
	Should flooding be shown by changing the hue from blue to red as danger increases?

Cartographic Communication

GI and maps are collected and ultimately made to communicate. How this communication works is important to think about because this reflection helps improve the GI and maps that we make. A more detailed consideration of the underlying issues follows in other chapters. The clearest communication results from the successful consideration of the many aspects of geographic and cartographic representation in the context of the purpose and any and all relevant disciplinary or professional conventions. After all, how the content can be communicated depends on geographic and cartographic representation choices. If you only count the list of choices made so far in this chapter (14), this means a total of 195 choices, which is still barely scratching the surface of the infinite potential in map design.

Before looking at the practical demands of GI and cartographic communication, we should consider an example of the relationship between geographic and cartographic representation. A common example occurs between the coordinate system and the projection. In many cases, the cartographic projection is the same as the projection used for the coordinate system. In these cases, it's clear how the relationship between the geographic representation and the cartographic representation makes it easier for cartographic communication. Staying with one projection means no transformation between projections is required, which limits the potential for distortions to be introduced. We need to consider how when the geographic representation and the cartographic representation use different projections, the potential distortions to areas, angles, shapes, distances, and directions can lead to many small errors or quite large errors. Combining the GI for monitoring river flooding in northern Illinois will be much more difficult if different projections and coordinate systems are used.

In summary, geographic and cartographic representation should always be considered together when examining or developing GI or maps for cartographic communication. Although the media for GI and maps are vastly different and the types of communication vary greatly, the parallels are great enough that with some exceptions the basic issues for cartographic communication apply to both.

Cartographic communication, in the most general sense, relies on distortions. As Mark Monmonier (1991) writes:

> A good map tells a multitude of little white lies; it suppresses truth to help the user see what needs to be seen. But the value of a map depends on how well its generalized geometry and generalized content reflect a chosen aspect of reality. (p. 25)

Although some cartographers find this view dismissive of cartographers' expertise and skills, it highlights how distortions are necessary for cartographic communication to succeed. Perhaps we should add the important note that this also means that we need to be especially on guard when working with material prepared under a cartographic license that is ambiguous.

For cartographic communication, we should pay special attention to the following characteristics of maps:

Scale	What is the relationship between units on the map and the same units on the ground? What simplifications accompany the scale?
Projection	What kind of distortion does the projection introduce to areas, angles, shapes, distances, and directions?
Symbolization	How do symbols exaggerate or minimize features on the map? How does the cartographic communication benefit from the chosen symbols? What is the best measurement framework? Nominal, ordinal, interval, ratio?
Generalization	How have irrelevant details for the map's purpose been filtered out? How have details relevant to a map's purpose been emphasized? How have lines, points, areas, and content been handled?

Conventions

Conventions are common and important for the reasons that have been highlighted earlier. The number of choices available—and the concerns that go along with them—may seem a barrier for activities involving GI and maps, but due to established conventions it is actually quite easy to read and create GI and maps. In most cases, conventions have already dealt with issues long before you have begun to read or create a map. Pragmatically, representation and communication distinguishes three types of conventions:

- Things most people anywhere in the world know
- Cultural influences and culturally influenced knowledge
- Disciplinary or professional understanding and knowledge

For example, most adults comprehend that water is symbolically represented by the color blue (the color of the Great Lakes in Plate 2). Even when you look at a map with text you cannot read, you will probably be able to distinguish water areas based on their color. However, examples like this are very, very rare. Culture exerts a powerful influence on how most people understand colors and conventions. For example, the color red, which for most people in Western countries symbolizes danger (e.g., traffic lights and fire), is the color for success in Chinese culture. The greatest number of conventions, however, come from disciplinary or professional subcultures. Disciplinary or professional groups have often developed complicated formal and informal codes for representation. Sometimes the symbols become ubiquitous through use—for example, interstate highway symbols in the United States—but many remain specific to disciplines—for example, pipe-line symbols used by sanitary engineers and field crews. Often effective GI and map "reading" and creation go hand-in-hand with an introduction to these conventions. In some countries—for example, Great Britain, South Africa, and Switzerland—children in school learn about their country's topographic map symbols and have little problem throughout the rest of their lives turning to the detailed maps made in these countries to orient themselves.

Scale and Accuracy

A key component of cartographic communication is scale. Usually scale reflects established conventions, rules, or possibly even laws. It has important consequences for accuracy. Scale can serve as a proxy for more complicated representations between a geographic representation and the phenomena on which it is based. Because of scale's significance for data collection and processing, indicating the scale for GI provides crude but effective shorthand for understanding the accuracy of GI. Large-scale maps cover small areas with significant detail and accuracy. Medium-scale maps cover larger areas with less detail and accuracy. Small-scale maps cover large areas with the least detail and accuracy.

The USGS describes the positional accuracy of their topographic products in terms of the U.S. National Map Accuracy Standard (1947) that states 90% of the points tested should fall within a fixed distance (0.02 inch or 0.5 mm) of their correct position. The points tested are only the well-defined points, which leaves open the possibility that less well defined points are far less accurate.

In terms of positional accuracy, it would be too easy to simply say that an accuracy standard means you know what you're getting. Remember that the actual location of these points could be anywhere within the areas indicated. In this sense, while the positional accuracy standard is a quantitative measure, its interpretation is often very qualitative unless exhaustive measure is made.

Scale also impacts attribute accuracy: fewer detailed measurements and observations can be made and represented at smaller scales. Smaller scales (larger areas) must generalize and combine phenomena into cartographic features that remain visible at the output or data analysis scale. Determining attribute accuracy is more complicated and generally involves comparisons of selected points on one map or in one GI data set with another or field checks at the actual locations. These tests are very important as they often indicate important differences between pieces of GI that are supposedly of the same phenomena, but in actuality are collected and/or analyzed using very different methods.

Quality and Choices

As discussed in Chapter 1, a common way to think about the reliability of a geographic or cartographic representation is in terms of quality. The highway map is "good" if we can use it to find our way easily. Simply said, good-quality GI or maps are useful for the purpose we create or intend to use them for. By "quality," we usually mean "reliability," but quality often simply means that a map fits the intended use. For example, you could use highway maps to figure out the average size of towns in a state, but the results would be of low quality. The concept "fitness-for-use" helps people working with maps and GI to get a grip on the slippery concept of quality. What this means is that we need to know the intended purposes of a map before we can decide what level of quality the map has.

It's important to recognize that the choices made for geographic and cartographic representation affect quality. The example of the differences between projections and coordinate systems is merely the tip of an iceberg that following chapters will develop in much greater detail. Going back to the river flooding example and thinking about the choices between vector and raster representations with their different levels of positional accuracy, you may grasp the significant consequences of this choice for cartographic representation and the accurate communication of flood events and how this is important for any and all GI and maps.

Summary

The representation of things and events from the world involves choices, which are greatly influenced by conventions. The choices are endless, making conventions critical to successful communication. Representation of things and events distinguishes between geographic and cartographic representation. Geographic representation is the abstraction of measurements and observations to GI. Sinton's framework provides a useful tool for considering different ways that things and events are geographically represented as information in terms of time, space, and characteristics (attributes). The relationships among things and events in a geographical representation are critical for cartographic representations. Cartographic representation creates maps and other visual representations and takes myriad cartographic presentation issues into consideration including scale, symbols, and graphic variables. Successful communication can be considered in terms of quality, especially the "fitness-for-use" of the GI or map.

REVIEW QUESTIONS

1. What is the difference between geographic and cartographic representation?

2. What are the four types of measurements?

3. How do geographic representation choices determine cartographic representation?

4. What distinguishes vector from raster data in Sinton's framework?

5. Why are TIN and topology not included in Sinton's framework?

6. What is the general relationship between scale and accuracy?

7. What does "cartographic communication" refer to?

8. What do databases, in an abstract sense, contain?

9. What is controlled in a stream gauge (using Sinton's concept)?

10. How is accuracy a qualitative indicator?

ANSWERS

1. What is the difference between geographic and cartographic representation?

   In essence, geographic representation is the selection of observations, measurements, and choices about their coding as attributes and relationships in a database. Cartographic representation is the selection of graphical elements and abstraction of GI to communicate for a purpose (or multiple purposes).

2. What are the four types of measurements?

   Following the psychologist Stevens, they are nominal, interval, ordinal, and ratio. Important measurements, including radial measurements, are not included in this widely used scheme.

3. How do geographic representation choices determine cartographic representation?

   How data is collected and stored as part of the geographic representation process limits the possibilities for making cartographic representations. For example, a multilane highway represented as a single line can never be used to indicate how many lanes of traffic are slowed down by heavy traffic.

4. What distinguishes vector from raster data in Sinton's framework?

   Vector data measures space and controls the attribute; raster data controls space and measures the attribute. Both types of data fix time for the geographic representation.

5. Why are TIN and topology not included in Sinton's framework?

   TIN and topology represent relationships. Sinton's framework involves space, time, and attributes. Relationships are at most implicit.

6. What is the general relationship between scale and accuracy?

   The larger the scale (the smaller the area a map represents), the greater the accuracy can be. The smaller the scale (the larger the area a map represents), the lower the accuracy can be. The main issue here is the possible size of the printed page. More detail can be fit onto a single sheet of paper at a large scale than at a small scale. This does not apply to GI. But because of the continued use of data collection processes used originally for producing printed maps, scale continues to offer a useful shorthand for assessing accuracy.

7. What does "cartographic communication" refer to?

GI and maps ultimately communicate. How they communicate depends on the cartographic representation choices. "Cartographic communication" refers to the ability of GI or maps to communicate for specific purposes.

8. What do databases, in an abstract sense, contain?

Attributes and relationships used for cartographic representation and communication.

9. What is controlled in a stream gauge (using Sinton's concept)?

A stream gauge controls time, measures attribute, and fixes space.

10. How is accuracy a qualitative indicator?

Some parts of accuracy may be quantitative, but assessments of accuracy also depend on the consideration of the potential use of the GI or map, which may be only partially specified, leading to a qualitative assessment of accuracy.

Chapter Readings

Chrisman, N. R. (1997). *Exploring geographic information systems.* New York: Wiley.

Gould, P. (1985). *The geographer at work.* London: Routledge.

Gould, P. (1989). *Becoming a geographer.* Syracuse, NY: Syracuse University Press.

Hartshorne, R. (1939/1956). *The nature of geography: A critical survey of current thought in the light of the past.* Lancaster, PA: Association of American Geographers. (Reprinted with corrections, 1961)

Hartshorne, R. (1958). The concept of geography as a science of space, from Kant and Humboldt to Hettner. *Annals of the Association of American Geographers, 48*(2), 97–108.

Monmonier, M. (1991). *How to lie with maps.* Chicago: University of Chicago Press.

Monmonier, M. (1993). *Mapping it out: Expository cartography for the humanities and social sciences.* Chicago: University of Chicago Press.

Monmonier, M. (1994). Spatial resolution, hazardous waste siting, and freedom of information. *Statistical Computing and Statistical Graphics Newsletter, 5*(1), 9–11.

Monmonier, M. (1995). *Drawing the line: Tales of maps and cartocontroversy.* New York: Holt.

Web Resources

↻ Geographic representation is an important topic for research in GIScience. See *www.spatial.maine.edu/~max/UCGIS-Rep.pdf.*

↻ A prominent GIS software producer's view on geographic representation can be found at *www.esri.com/industries/k-12/education/~/media/files/pdfs/industries/k-12/pdfs/geoginquiry.pdf.*

↻ The application of cartographic representation to mapping in Africa offers important insights into pragmatic issues. See *www.africover.org/carto_standard.htm.*

↻ A discussion about changes to cartographic representation helps one to think about connections to geographic representation. See *www.questia.com/library/1G1-72433261/representation-and-its-relationship-with-cartographic.*

↻ The U.S. EPA (Environmental Protection Agency) has some interesting materials on the differences between raster and vector at *www.epa.gov/region02/gis/gisconcepts.htm.*

↻ These presentation materials provide a succinct overview of accuracy issues: *www.epa.gov/nerlesd1/gqc/courses/images/kirkland.pdf.*

EXERCISES

1. Geographic Representations: Measurements, Observations, Relationships

Think of some environmental or social issues of which you are aware and consider how to represent them geographically. Take into account how you would measure the observations and relate the measurements (as geographic information) to each other. What measurement types could you use for the observations?

2. Geographic Representations: Considering Choices

Starting with the questions provided in the list of choices, consider some of the potential issues in making the geographic representation you started in Exercise 1 above. Any issues are valid in this exercise. Write them down and see how they show themselves in the following chapters.

3. Quality and Choices

Based on the discussion of geographic and cartographic representation and how they relate to cartographic communication, discuss the impact of your choices on quality in Exercises 1 and 2 above.

4. **Representations: Identify Different Measurements and Geographic Representations**

Overview

A key part of creating and working with GI and maps is identifying the underlying measurements and choices in geographic representations. This can be difficult, but extremely worthwhile.

Instructions

Using maps from the library or Internet sites (e.g., *www.davidrumsey.com*), identify at least two different thematic maps. Examine the maps and identify the measurements, geographic representations, and cartographic representations.

Questions

1. What is the title, subject, and date of the map? Who created the map? Is it part of an atlas, series, or report? Where did you find the map?

2. What alternative measurements could have been made? Explain at least two measurements and the consequences for the map.

3. Is it clear how the measurements are collected? Does it say when and by whom? How is this information (or lack thereof) important for assessing the geographic and cartographic representations?

CHAPTER 3

GI and Cartography Issues

"All maps lie" is a statement that reflects the necessity of relying on abstractions in order to communicate with maps. After considering GI and maps in terms of geographic representation, cartographic representation, accuracy, and quality in the previous chapters, it's clear that any map's "lies" are always matters of dispute. In some sense all maps are the results of particular choices and conventions concerning how to abstract. Some maps intentionally and overtly distort and hold little similarity with the things and events in the world they represent. For instance, advertising maps often artistically collapse distances. But the abstractions can be far more complex. Other maps, by simplistically applying conventions and frameworks, using a Mercator map to compare per capita GDP (gross domestic product) of countries, for example, may greatly but unintentionally distort the representation of the world by suggesting some areas are much larger in comparison to other areas of the world. Still, even these distorted maps and GI are useful, as they often are the only way to see and get to know parts of the world, or to see the distribution of population over the entire world. Choices made in the geographic and cartographic representation of GI and maps determine accuracy and quality. Whether there is "too much" distortion comes down to a map's fitness for the use it is put to.

This chapter introduces the fundamental choices of geographic and cartographic representations covering the key properties of projections, scale, symbolization, and color. It reviews concepts independently of more detailed examinations of cartographic and GI principles in the following parts and later chapters. This chapter provides an orientation in the context of choices with the goal of deepening the introductory material of Part 1 to gain a better understanding of issues in creating and using GI and maps to understand and communicate about the world. These concepts are especially important to the successful creation and use of maps and GIS.

Examples in Chapter 4 explore some actual cases and issues and provide more specific cases.

From a Round to a Flat Surface: Projections

Representing and communicating geographic information and maps on flat surfaces (screens or paper) requires transforming three-dimensional locations from the earth to two-dimensional locations on a flat plane (see Figure 3.1), which can be either a two-dimensional (2-D) coordinate system (in the case of GI) or a piece of paper (in the case of a map). Projections for large portions of the earth usually use a simple sphere; for smaller areas, where positional accuracy gains in importance, the projection uses an ellipsoid that locally corresponds to a **geoid**, the name for the shape with the most accurate correspondence to the actual oblate and irregular shape of the earth at a given time.

Projections have great importance for GI because most GI records the location of things and events on a two-dimensional coordinate system, called a **Cartesian coordinate system** when the x and y axes intersect at right angles (see the right side of Figure 3.1). Polar coordinates, which record location in terms of one distance and an angle from a central point, are also used in some applications. Any projection of location from the round surface to a flat plane causes some form of distortion. This has important consequences for the accuracy of GI or maps and what you can do with a particular projection.

Traditionally, most books on cartography start by discussing projections. Projections are one of cartography's most important contributions to science and civilization. Projections are, and have been, the foundations for almost all representations of the earth or any part of the earth. Almost all GI also uses projections. The ancient Greek geographer and astronomer Ptolemy invented several projections that were used by the Romans and by others

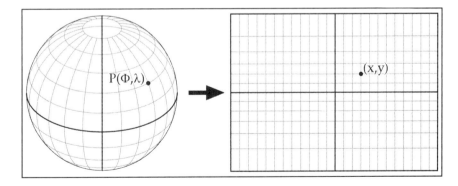

FIGURE 3.1. Most projections of locations transform three-dimensional locations into two-dimensional locations as a geographic representation and basis for cartographic representation.

for centuries afterward. Some centuries later, when European exploration and colonization commenced, because they were so important to the accurate determination of a ship's location and showing geographical relationships between mother countries and colonies, projections quickly became an important mathematical activity. You can even think of the 400 years between Mercator's publication of his global projection (see Figure 3.2) in 1568 and 1968 as the "golden" years of projections (see Figure 3.3). Although the choice of these years is somewhat arbitrary, it roughly begins with the period of significant European colonization and ends soon after computers made the calculations for projections a much easier task. Before moving on to the concepts of projections, you should also know that while it is possible to record the location of things and events in three-dimensional coordinate systems, they are still rather uncommon in most of geography and cartography. They are very uncommon because of their relative complexity, the widespread use of two-dimensional coordinate systems, and the cost of transforming two-dimensional coordinate systems. Chapter 5 takes a look at some of these systems, including their applications, in much greater detail.

Key Concepts of Projections

Projections convert measured locations of things and events in three dimensions to two dimensions. Projections are important but also complicated because it is impossible using geometric or more complex mathematical

FIGURE 3.2. A world map from 1801 using a Mercator projection is just one example of many from the last 400 years.

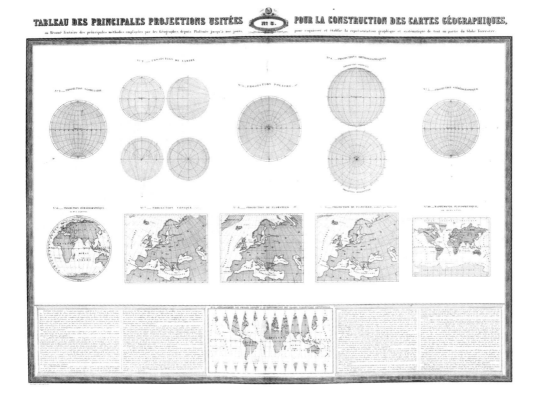

FIGURE 3.3. Illustration from 1862 showing 15 projections for different areas of the earth. The basic typology is still helpful today in choosing an appropriate projection for a desired geographic representation and cartographic representation.

methods to simultaneously preserve both the shape and the area of any object found either on the three-dimensional spherical surface of the earth, in the earth, or near the earth, when we depict it in a two-dimensional coordinate system. Each projection is an abstraction of the earth's surface and introduces distortions that affect the accuracy of the GI or map. A projection starts with one of three representations of the earth's irregular surface (geoid, ellipsoid, or spheroid) and converts it directly or through intermediary transformations to a flat, or planar, transformation (see Figure 3.4).

Choosing the right projection is important for controlling these distortions. Thankfully, choosing the right projection for a particular area is a task that has often been done by institutions and governments and made part of conventions or even laws that define the projection and its parameters that must be used for certain areas and activities (see Chapter 5). This is usually a good thing, but many institutions and governments require multiple projections. Some confusion can result in choosing the wrong projection (see Chapters 4 and 12).

Whatever you do with GI or maps, you need to know some projection concepts in order to understand projection distortions and their

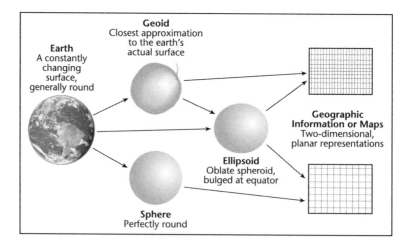

FIGURE 3.4. Some fundamental abstractions of the earth used in choosing projections for a desired geographic representation and cartographic representation.

consequences. Some GI is stored in latitude and longitude coordinates and can be displayed or mapped on a flat screen or piece of paper, but these "unprojected geographic coordinates," as they are usually called, have tremendous amounts of distortion when shown on a flat plane.

Four fundamental concepts are crucial to know when you use GI and maps:

1. *The earth is almost round and always changing shape.* Three models of the earth are used in making projections: sphere, ellipsoid, and geoid. A perfectly round object, or *sphere*, is defined by the mathematical relationship between the center of the object and its surface, the radius. The surface of a sphere is a constant distance from the object's center. This is the simplest model used in projections and is sufficient for GI and maps of very large areas. However, because the spinning of the earth creates a centrifugal force that causes the earth to bulge at the equator and flatten at the poles, the distance from the center of the earth to any point on the equator is greater than the distance between the center of the earth and the north or south poles. This more precise shape is known as an **ellipsoid** (but often called a **spheroid**) and comes much closer to describing the actual shape of the earth. It is accurate enough for most GI and maps of smaller areas. Because of different weights of material in the earth's core, differences in magnetic fields, and movements of the earth's tectonic plates, very detailed measurements of locations use a geoid for projections. A *geoid* is the most accurate representation of the earth's surface. It accurately describes the location of objects to a common reference point, usually for only relatively small areas. Differences in the location of an object among the sphere, ellipsoid, and geoid can be as much as several hundred meters (yards). The ellipsoid and geoid models of the earth are defined and updated at irregular intervals. Should you become involved with very detailed and accurate measurements

of location, you should also be aware that the geoid of the earth is constantly changing and locations recorded with an older geoid may not match a newer geoid. (See Plate 3 for geoid undulations.)

2. *A projection makes compromises.* Every projection either preserves one projection property or makes some compromises between projection properties (see Figure 3.5). In either case, some projection properties are compromised by every projection. Because there are theoretically an unlimited number of projections, it is important to organize projections by projection properties. Which projection is used in making GI or a map has much to do with how geographic characteristics and relationships are preserved. The four projection properties, along with the cartographic terms in parentheses for each, are:

Angles	Preservation of the angles (including shapes) of small areas (conformal)
Areas	Preservation of the relative size of regions (equivalent or equal area)
Distance	Partial preservation of distance relationships (equidistant)
Direction	Certain lines of direction are preserved (azimuthal)

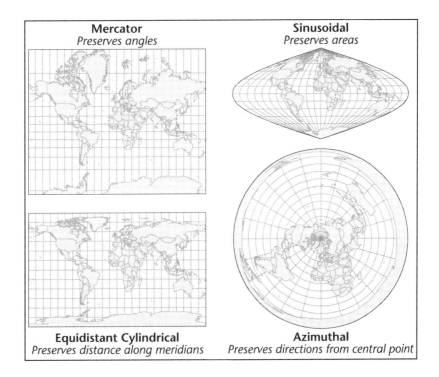

Mercator
Preserves angles

Sinusoidal
Preserves areas

Equidistant Cylindrical
Preserves distance along meridians

Azimuthal
Preserves directions from central point

FIGURE 3.5. Four basic types of projections with their projection properties that impact geographic representation and cartographic representation.

Most projections preserve area, although a large number are **compromise projections**, which means that they sometimes preserve area and sometimes preserve shape. Usually compromise projections are used for showing the globe, but they can be used for smaller areas. All things considered, the projections that preserve area are more common because people usually need maps of smaller areas where geographic relationships and area comparisons are very important. However, the projections showing the globe are significant because they are the only way for almost all people to see and understand the world. Global projections make very significant trade-offs between projection properties. One of the most common projections used for showing the entire world, the Mercator projection, is a classic case of how a projection always trades off different projection properties. In the case of the Mercator projection, it preserves the shape and distance relationships of small areas, but only locally; it preserves lines of constant bearing; it fails to preserve area (the sizes of Greenland and Africa are greatly distorted); it partially preserves continuity, breaking Eurasia into two halves. These trade-offs mean that the Mercator projection is a good choice for representing small areas and large areas, but only for navigation.

3. *Distortions will occur.* Every projection, in making trade-offs between the various projection properties, creates distortions. These distortions can be minimized by choosing a projection that corresponds as well as possible to characteristics of the area to be mapped and the known purposes and uses of the GI or map. Inappropriate and erroneous choice of projections can lead to significant errors and misrepresentations. Since there are no rules for choosing optimal projections, you simply have to assess each projection individually and learn through practice and discussion with other people what projection is best for a particular area, purpose, and use. In many places the projections of most GI and maps have already been determined. However, different people, institutions, and countries may use very different projections for the same area, requiring you to know the distortions that different projections create (see Figure 3.6).

4. *GI from different projections should not be combined.* GI is particularly prone to errors resulting from the combination of data from different projections. This also applies to maps, but since it is very time-consuming to trace two maps and overlay the tracings, in practice you should be most concerned with the consequences of combining GI from different projections, which is perhaps one of the easiest mistakes to make with GIS. Sometimes, although you may know the GI is for the same place, the combined data is separated by a huge distance, possibly even many times the size of the earth. Sometimes—and this is why knowing the projection of GI is so important—the distances between GI objects can be minute, just a few inches or feet. However, because of differences in projections, what may be minute differences in one place may be vast differences elsewhere.

Assessing projection distortions and determining the best projection for an activity and area remains a complex activity that can be required for working with very accurate geographic information. Metadata often provide crucial information about the projection of GI. (See Chapter 5.)

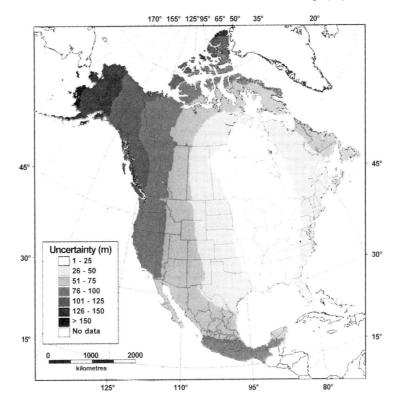

FIGURE 3.6. Positional uncertainty when coordinate system datum is unknown in North America (NAD27 or NAD83).

Projected or Unprojected GI

GI or maps for large areas—for example, a continent or the world—are often projected, but they can also be unprojected. If they are unprojected, the distortion is very significant. In many cases, the latitude and longitude values are converted to a two-dimensional orthogonal network of *x, y* values. The advantage is that unprojected GI can readily be transformed to other projections as needed. The data or maps of smaller areas are usually projected because the projected representations better correspond to conventional maps that people have used for many years. In areas with legally established coordinate systems or with clear conventions, the choice of projection can be easy and prescribed (see Figure 3.7 for an example of legally established projections from the United States). In other areas, a few choices may be preferable depending on the orientation, size, and accepted practices for the area in question.

Projections in Practice

You should look at the distortions of the Mercator projection and the recently popularized Peters projection in Figure 3.8. The widespread use of

FIGURE 3.7. U.S. continental State–Plane Zones (NAD83). These zones are commonly used in the United States for state geographic information activities and their use can be required by state or local statute.

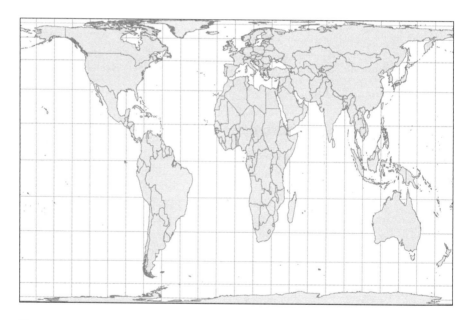

FIGURE 3.8. The Peters projection is very similar to this Gall projection. A compromise projection, its often praised for its suitability for global mapping, but in fact it distorts both areas and shapes.

the Mercator projection to show things and events at a global scale (which, you should note, Mercator never did and cartographers advise against) leads to very sizable distortions, especially in areas near the poles, but also in the latitudes where most of Europe and North America are located (see Figure 3.9). These distortions led Arno Peters to promote his adaptation of older projections, the Peters projection, which has been widely adopted even though it introduces other distortions. While the Peters projection does not solve projection problems, it has made people more aware of the distortions inherent in projections.

GI and Maps Are Abstractions

Finally, we should note that projection is one type of abstraction, which can be misused and even lied with. Sometimes this is obvious, but careful editing can gloss over rough spots. GI and maps involve many other abstractions, which is why one of Mark Monmonnier's books on cartographic principles, uses, and abuses carries the title "How to Lie with Maps." Based on what you now know about projections, the claim that maps lie is easy enough to refute. All maps must have distortions; therefore, some would argue, what is called a "lie" is only a "distortion."

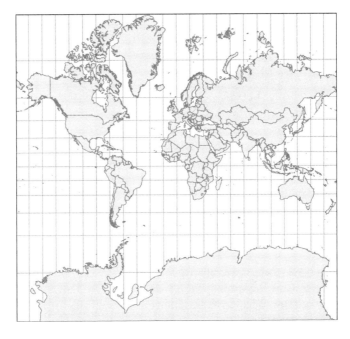

FIGURE 3.9. A Mercator projection showing extremely distorted areas near the north and south poles. Distortions can be desirable, but generally they promote ill-suited cartographic representations.

Additional Fundamental Choices

The principal choices in geographic and cartographic representation highlight key issues for the GI or map framework that abstracts the infinite complexity of the world. This process of abstraction is commonly called "data modeling" in GIS practice and teaching. Data modeling merges geographic and cartographic representation. However, when using existing GI, other organizations have already decided many aspects of data modeling. By keeping geographic and cartographic representation separate, you can develop a better understanding of the complexities of data modeling.

Things or Events

One of the most primary choices is whether to depict the phenomena of interest as things or as events. Usually, the geographic representation of things is easier. However, showing phenomena as things makes it necessary to make more complicated cartographic representations later that show the relationships among things. For example, showing a traffic jam with a symbol generally suffices to show where the traffic is stopped or slowed down. How the traffic jam develops, however, cannot be easily shown with this geographic representation. Modeled as an event, possibly at the level of individual cars and trucks, the development of the traffic jam can be represented as a dynamic process.

Patterns or Processes

The choice between patterns or processes is inseparable from the geographic representation choice of things or events. Still, while it is impossible to represent a process solely relying on things, it is possible to represent events as patterns, or to add additional GI information to the things to show more of the process. The addition of GI to a geographic representation to support cartographic communication objectives—for example, how a detour for additional traffic decreased the size of the traffic jam—is a possibility for addressing some of these issues.

Abstracted or Accurate

All GI and maps are abstracted, but choices remain in this regard in attempting to find the appropriate balance between abstraction and accuracy. The abstraction of a cartographic representation can make the communication of important relationships easier, but diminished accuracy can make it difficult to use the resulting GI or map for other activities.

Few or Many Associations

Along with a balance between abstraction and accuracy, the number of associations in the geographic representation and the cartographic representation opens up some challenging issues. For communication, a map reader or GI user has to be able to associate the final graphical product with his or her own experiences and knowledge. Sometimes, for example, in schematic maps of utility lines, the contextual information of a place is kept very simple and the components of the utility network are almost the only things appearing in the cartographic representation. The traditional topographic map goes in the other direction, offering a multitude of cartographic elements that can be associated in myriad ways with a person's experiences and knowledge. Most cartographic representations lie somewhere in between. Simplifying the associations can be effective in focusing the representation on specifics, but runs the risk of losing critical details necessary for using the GI and maps. Coming back to the traffic jam example, a detailed breakdown of the traffic jam length and average speeds in individual lanes may be useful for showing relationships between entering traffic and the traffic jam, but could add too much detail for a communication goal of showing simply where the traffic jam is and how long it lasted.

Scale

The *scale,* or the relationship between a unit of distance on the screen, the size of a collection unit, or the units of distance used on a map to the same unit of distance on the ground, represents a critical choice. If the scale is large (showing a small area), a great amount of detail can be represented. If the scale is small (showing a large area), then less detail can be shown. (Chapter 5 presents more information on scale.) This relation of scale to detail and area constrains maps greatly and has impacts on GI. Although computers allow for zooming in at different scales to data, data captured at a small scale becomes very inaccurate when it is zoomed in to. What scale is chosen, whether for GI or maps, impacts both the geographic representation and the cartographic representation. (See Plate 4 for graphic symbols used in cartographic representation.)

Symbolization

A fundamental choice for cartographic representation is the symbolization. **Semiotics**, the study of signs, helps us to understand the meaning of symbols and how symbols take on meaning, both individually and through relationships with other symbols. Significant choices for cartographic representation involve symbols.

At a fundamental level, the choice of cartographic symbols and semiotics can be compared to a language. In the "cartographic language," a limited set of graphic variables is available for "writing" a map. Building on Bertin's earlier work on general graphic variables, several cartographers identified size, pattern, shape, color value, color hue, color saturation, texture, orientation, arrangement, and focus. The graphic variables can be combined in myriad ways, but it is clear that some variables are more associated with difference in quantities—for example, size—than others. Effective cartographic communication depends on how well the map creator matches graphic variables to the spatial dimensions of the things, events, and associations.

Through semiotics, both the cartographic creator and the reader can assess the connection between the symbols and the represented geographic things and events. The Minnesota Department of Transportation depicts traffic volumes in the Twin Cities using the three colors green, yellow, and red (see Plate 1). The system is extremely effective in rapidly communicating traffic slowdowns. In spite of the complicated highway network, because conventional colors for traffic signals are used, people can quickly understand where traffic is slowed down or stopped. This is a meaningful use of a graphic variable to communicate a complex thing. (See Chapter 11 for more discussion about cartographic representation.)

Geographic representation involves choices with direct impacts on symbols and semiotics. Whether it be traffic jams, soil pH, water flow, dispersion of airborne pollutants, or household income, how the data is geographically represented plays a huge role in what symbols can be used and how the meaning and significance of the meaning can be communicated.

Color and Symbolization

The use of color in cartographic representation involves several important choices. Value, hue, and saturation are graphical characteristics of color (see Plate 5), which are most significant for pragmatic purposes (see Chapter 11 for further discussion). **Color value** refers to the different degrees of darkness or lightness of a color. High values are light and low values are dark. Color value is usually applied to distinguish ordinal data values—for example, soil pH or population density. **Color hue** is what people normally refer to as "color," which is the distinction between blue, brown, red, yellow, and so on. This distinction is a result of the reflectance of different light wavelengths by a surface. The ability to distinguish hue is commonplace among people in all cultures, but the significance of individual colors can vary widely. Color hue can be used to show nominal differences—for example, different states, types of vegetation, planning zones—but should be used very carefully for numerical values because it is difficult for people to associate a large number of hues with changes in values. *Color saturation* is the purity or intensity of a hue. Saturation is used in conjunction with value and hue to enhance reader perception of relationships and order of map features.

Organizational Structure of GIS, Software, Hardware, and Peripherals

Organizations, software, hardware, and peripherals have great influences on the range of choices available for geographic and cartographic representation (see Figure 3.10). Nowadays the various aspects of geographic representation, cartographic representation, conventions, and choices come together in the organization of GIS. A GIS always has an organizational

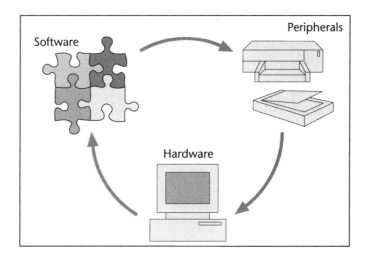

FIGURE 3.10. GIS organization as an arrangement of hardware, software, and peripherals.

aspect: people obtain data from other people or organizations, share the results of their analysis or mapping with other people, and coordinate the work related to GIS with other people or organizations. Additional key aspects of the organizational aspects of GIS are covered later in Chapter 12.

Key computer components of GIS organization are software, hardware, and peripherals. GIS *software* contains the programs, interfaces, and even procedures for processing data and making graphics and maps. *Hardware* performs the operations of the software and makes displays and printouts. The *peripherals* can be part of the core computer hardware (CPU and memory) necessary for the basic GIS software operations, but can include printers, tape drives and external hard disks, monitors, digitizers, scanners, and so on, that offer additional possibilities for working with the GIS.

Because of the complexity of the computing operations, GIS often requires supporting organizations. Even a single person working with GIS will want, or even need, help with the hardware, software, and peripherals from time to time. The size of the GIS organization often goes hand-in-hand with the size of the company or office where the GIS is being used. There certainly is such a thing as desktop GIS, but no matter how good the marketing, the complexity of GIS necessitates good support. Unfortunately, the people working with and using a GIS are often overlooked when considering acquiring a GIS.

Beyond people, who are often not depicted in representations of GIS organization, GIS also rely on measurements and conventions to a high degree. These aspects, often abstract and implicit, are key parts of the underlying framework for all work with GI and maps. Of course, without data, all the hardware, software, and peripherals will never make a GIS.

Summary

The choices made for geographic and cartographic representation determine accuracy and quality. Fundamental choices involve projection, scale, and symbolization. A projection is used for all accurate GI and maps because the coordinates of almost all GI are recorded in a two-dimensional coordinate system; however, the earth is a three-dimensional object. It is possible to calculate locations for GI or maps without using a projection, but positional accuracy is lost. Other choices involving scale and symbolization are fundamental and are often included in the data modeling that goes along with the geographic and cartographic representation. Scale is important, for it determines the area and detail of GI and maps. Symbolization leads to the meaning of symbols that should assure accuracy and quality. Color is one of the most important symbolization choices. Its use often follows established conventions. Beyond conventions, the organization of GIS including software, hardware, and peripherals can have a very strong influence on the choices that can be made.

REVIEW QUESTIONS

1. What GIS component is frequently overlooked in descriptions of GIS?

2. What is the difference between a *spheroid* and an *ellipsoid*?

3. How is a geoid used in relation to an ellipsoid?

4. How is scale helpful for working with GI?

5. Why would some people claim all maps lie?

6. What are the four main choices in creating GI?

7. What are the three components of color in the HSV color model?

8. What is the most common type of projection used in surveying?

9. What projection properties does a Mercator projection preserve?

10. What is the difference between large and small scale?

ANSWERS

1. What GIS component is frequently overlooked in descriptions of GIS?
   People are often overlooked.

2. What is the difference between a *spheroid* and an *ellipsoid*?
   A *spheroid* is perfectly round, an *ellipsoid* is oblate.

3. How is a geoid used in relation to an ellipsoid?
   An ellipsoid can be "fitted" to a geoid.

4. How is scale helpful for working with GI?

   Scale helps in understanding the abstractions between collection units and mapping units, as well as in understanding characteristics of the GI representation.

5. Why would some people claim all maps lie?

   All maps and GI must abstract from the actual things and events in the world, thus creating distortions.

6. What are the four main choices in creating GI?

   The four main choices are projection, scale, data, and symbols.

7. What are three components of color in the HSV color model?

   The three components of color are hue, value, and saturation.

8. What is the most common type of projection used in surveying?

   The most common type of projection used in surveying is conformal, which preserves angles.

9. What projection properties does a Mercator projection preserve?

   A Mercator projection preserves direction.

10. What is the difference between large and small scale?

    Small scale shows large areas, large scale shows small areas.

Chapter Readings

Bertin, J. (1983). *Semiology of graphics: Diagrams, networks, maps.* Madison: University of Wisconsin Press.

Chrisman, N. R. (1999). What does "GIS" mean? *Transactions in GIS, 3*(2), 175–186.

Clarke, K. C. (1995). *Analytical and computer cartography.* Upper Saddle River, NJ: Prentice Hall.

Cotter, C. H. (1966). *The astronomical and mathematical foundations of geography.* New York: Elsevier.

Dent, B. D. (1993). *Cartography: Thematic map design.* Dubuque, IA: Brown.

MacEachren, A. M. (1994). *Some truth with maps: A primer on symbolization and design.* Washington, DC: American Association of Geographers.

MacEachren, A. M. (1995). *How maps work: Representation, visualization, design.* New York: Guilford Press.

Morrison, J. L. (1994). Cartography and organization. *ACSM Bulletin, 148,* 38–45.

Pickles, J. (2004). *A history of spaces: Cartographic reason, mapping, and the geo-coded world.* New York: Routledge.

Robinson, A. H., & Snyder, J. P. (Eds.). (1991). *Matching the map projection to the need.* Bethesda, MD: American Congress on Surveying and Mapping.

Suchan, T., & Brewer, C. A. (2000). Qualitative methods for research on mapmaking and map use. *Professional Geography, 52*(1), 145–154.

Wieczorek, J., Guo, Q., & Hijmans, R. J. (2004). The point–radius method for georeferencing locality descriptions and calculating associated uncertainty. *International Journal of Geographical Information Science, 18*(8), 745–767.

Web Resources

🕑 A general overview of projections can be found at *http://geography.about.com/library/weekly/aa031599.htm*.

🕑 For some helpful information about map projection distortion, see *www.progonos. com/furuti/MapProj/Normal/CartProp/Distort/distort.html*.

🕑 For a very comprehensive set of graphics showing a great number of projections, see *http://egsc.usgs.gov/isb/pubs/MapProjections/projections.html*.

🕑 For more information on perhaps the most seminal work on projections, John Snyder's *Map Projections: A Working Manual*, see *http://pubs.er.usgs.gov/pubs/pp/pp1395*.

🕑 More mathematical descriptions of projection are available at *http://mathworld.wolfram.com/topics/MapProjections.html*.

🕑 For general information about scale, see *http://geography.about.com/cs/maps/a/mapscale.htm*.

🕑 The USGS provides general information about map scale at *http://erg.usgs.gov/isb/pubs/factsheets/fs01502.html*.

🕑 The importance of scale for geographic representation is handled at *http://historymatters.gmu.edu/mse/maps/question3.html*.

🕑 To see how cartographic generalization developed, read *www.ngs.noaa.gov/PUBS_LIB/Cartographic_Generalization_TR_NOS127_CGS12.pdf*.

EXERCISES

1. Principal Choices of Geographic and Cartographic Representations

This exercise has two parts that can be broken up into two separate exercises if time doesn't allow for them to be done together.

In the first part of this exercise, take a close look at a map and describe how the map shows principal choices of geographic and cartographic representation. Start out by defining the purpose, the scale, and the area of the map and then consider the principal choices described on pages 64–67 in the book. On a sheet of paper, make seven rows for the choices: things or events, patterns or processes, projected or unprojected, abstract or accurate, few or many associations, scale, and areas. In one column for each row, write a few words explaining the choices as you look through the map. After you've filled in all the choices, create another column and point out the relationships with other choices. In the second part of this exercise, together with a neighbor in class, prepare a description of the choices you would make to create a map for a purpose. It could be the way to a weekend sports event, the bus routes in your town, commuter maps, or the like. You should start out by defining the purpose, the scale, and the area of the map, and then follow the choices described on pages 64–67 in the book. On a sheet of paper, make seven rows for the choices, making sure to leave a place for you to show how a sample map will look: things or events, patterns or processes, projected or unprojected, abstract or accurate, few or many associations, scale, and areas. In one column for each row write a few words explaining your choice and in another column point out any relationships with other choices.

2. Detail the Fundamentals of Geography and Cartography in a Real-World GIS Application

Analyze an existing GIS application (national or global) and identify choice of projections, how GI is abstacted/derived from the data, scale choices, choice of symbols, and what GIS components are used: software, hardware, people, and organizations.

EXTENDED EXERCISES

3. Projections

Objectives:

Identify types of projections
Describe the properties preserved and sacrificed by each type of projection
Relate projection to different orientations of geographic areas

Overview

Projections are crucial for geographic and cartographic representations. You should be able to identify different types of projections and relate them to the projection properties they maintain and to the projection properties they compromise. Finally, you should also be able to distinguish between types of projections used for different orientations of geographic areas.

Instructions

Using maps from the library or Internet sites that show different types of map projections (e.g., *www.davidrumsey.com*), identify at least four different projections. Using information from this chapter on map projection properties, identify the projection properties preserved by the map projection and the map projection properties compromised by the projection. Compare the orientation of the areas shown on the different maps (north–south, east–west) and the projections used.

Questions

1. What types of projections did you identify? What are the names and subjects of the maps?

2. What projection properties do the maps maintain and what projection properties do the maps compromise? Make sure to clearly identify the names of the maps along with the names of the projection and properties.

3. Is there any correlation between the types of projections and the orientation of the geographic areas shown in the maps?

4. Fundamental Choices

Objective:

Working with maps from the library or the Internet, assess the choices made in the creation of the maps and the resulting consequences for geographic and cartographic representation.

Overview

The choices made in creating GI and maps determine what can be done with the GI and maps. Identify these choices as best you can and discuss alternatives for each choice.

Instructions

Using maps from the library or Internet sites that show different types of map projections (e.g., *www.davidrumsey.com*), identify two different thematic maps. Start out by describing the purpose, the scale, and the area of the map and then follow the choices described on pages 64–67 in the textbook. On a sheet of paper, make seven rows for the choices, making sure to leave a place for you to show how a sample map will look: things or events, patterns or processes, projected or unprojected, abstract or accurate, few or many associations, scale, and areas. In one column for each row write a few words describing the map's choice and in another column point out any relationships with other choices.

Questions

1. What are the most important choices made for each map? Describe each map and explain your reasoning.

2. What alternative choices could be made? Explain at least two choices and the consequences.

3. Did you identify any choices that lead to inaccuracies? If not, give an example of a choice for one of the maps you used for this exercise and present a choice that would lead to inaccuracies.

CHAPTER 4

Some History and the Continued Many Uses of GIS

Even if you've only heard of GIS recently, frankly it is quite difficult to imagine an aspect of human life without GIS-related impacts. GIS has become a cornerstone information technology for problem solving used whenever people have needs or interests related to location and geographical relationships. And this applies not just to situations where people used to use maps. Maps remain key media for GI because of their unparalleled flexibility, accessibility, and allure in any situation, even without reliable electrical power, but today many maps still are created with the help of GIS. With its greater analytical and representational flexibility, GIS helps to solve countless kinds of problems. GIS helps emergency responders to more quickly answer calls of distress, helps delivery companies provide more efficient routes, helps airplanes navigate more safely, and on and on. GIS provides important technologies for linking the capabilities of computers to the geographical knowledge and communication of civilization. The abilities that GIS provides have altered the ways people understand, communicate, and engage with the world around them and beyond them.

In every use, GIS, as a way of representing and communicating about things and events in the world, involves making certain choices and following conventions. Some things are missing, some things simplified, and some things exaggerated. And with GIS, even if it is not immediately obvious, there is always the issue of representing the round surface of the earth in a two-dimensional coordinate system through projections. We use GI to know and learn about things we may otherwise never experience. To communicate well, to help make good decisions, choices must be made, and following conventions can help.

This chapter briefly reviews the history of GIS and the importance of problem-solving uses, puts GIS into relationship with other technologies and institutional arrangements, and provides some application examples. The summaries of the examples, brief as they are, point to concepts and fundamentals of geographic data collection, cartographic representation, and geographical analysis in terms of the choices and conventions that determine how GIS is used.

A Brief History of What We Now Call GIS

GIS plays a central role in how we spatially enable activities, but what it has become has a long history. What we now call GIS was once known by different names, mainly maps, but also statistics, quantitative geography, data handling, information processing, and more. From all that and the first widespread uses of the term GIS in 1967, GIS has become more than the sum of its parts. When considering the infinite possibilities of GIS applications, you will almost always end up at some time with a statement of the principle that GIS helps with problem solving. Taken alone, this statement may be helpful for abstract discussions, but we need to have something more specific to refer to as a benefit, even if it is just a potential benefit. In the discussion of examples later in this chapter, you can read about specific examples and how choices and conventions impact what and how these examples deal with fundamental concepts.

Before getting there, a brief history of GIS shows that how people develop GIS has emphasized supporting problem solving; their choices and the conventions they follow are related to the fundamental concepts presented so far in this book.

The first ideas for using computers in geography go back to the 1940s, but the connection to spatial analysis and visualization starts to appear in the 1950s. Cartographers were creatures of habit, and beneficiaries of large institutions that invested in producing maps and related products. This meant the adoption of computer innovations moved slowly. Academics involved in planning and theoretical geography, William Warntz, Edgar Horwood, and Waldo Tobler among others, began then to think about and explore ways to use computers to make maps. Motivations for this endeavor were wide-ranging, some arose from the desire to make use of data that larger municipalities and governments were already collecting to fulfill legal obligations, some arose in the general excitement of the space race and the increasing presence of computing in visions of the future, some were due to mathematical advances and ideas of how to extend geography and other fields through mathematics and quantitative techniques, and finally some were due to the simple desire to reuse resources and avoid complicated and costly redrawing and recompilation of maps.

With the, relatively speaking, increase in availability of computers, more people saw the potential of computers and were intrigued by the possibilities of their use in making maps. A few academics also thought that the use of computers would help break through the status quo of cartography, even reforming cartography, and they actively sought ways to improve on the first computer maps. This innovation went hand-in-hand with the development of analytical uses of mapping for many disciplines, the first applications. In *Charting the Unknown* Nicholas Chrisman points out large projects studying urban air pollution in the late 1960s, and there are many examples of urban renewal related projects at state, county, and municipal levels in the United States. In many countries across the world, parallel developments took place, such that by 1972 Roger Tomlinson organized a symposium on geographical data handling that produced a capacious overview of the current developments. By this time a very powerful synergy that linked theoretical approaches to practical problem solving through innovative technological development had prepared the foundation for the blooming of GIS in the 1970s and 1980s.

From the mid-1970s on, GIS was rapidly growing and slowly becoming the ubiquitous system we know now to support problem solving and research in any domain or profession that deals with geography. Refinements and specific developments for the requirements of that growing number of applications became the dominant focus. By the mid-1990s the professional uses of GIS had grown so much that academics began to search for more academic pursuits, leading to Michael Goodchild's seminal proposal to develop geographic information science (GIScience). GIS and GIScience have since then been fruitfully connected, allowing for each to develop its strengths and draw on the other as needed. Indeed, now a huge economic sector, GIS turns frequently to GIScience.

Ongoing and Future Developments

How large of an economic sector is GIS now? A recent report (Oxera Consulting, 2012) concludes that global revenues in the geo services sector lie between $150 and $170 billion. In the United States, a report by the Boston Consulting Group (2012) points to a geo services sector with $73 billion in revenue with economic impacts 15 to 20 times that amount. Worldwide, the geo services sector produces a 0.2% portion of global GDP. Over 50% of Internet users access maps online, which, according to Oxera Consulting, contributes to higher-order thinking and impacts of $12 billion per year.

A trajectory showing the growth of GIS-related economic activities from 1967 to 2013 would most likely be one showing exponential growth during most of the period. By the time you read this book, the impact is obvious. GIS is everywhere, and now maps have gone from being an obligatory tool and medium for geography and cartography to becoming a consumer implement and artifact of life in the information society. Spatial data infrastructures (SDI) network GIS to spatially enable people. A life without maps may never happen, but the success of GIS means it increasingly recedes from the limelight and becomes part of business, government, and scientific infrastructures. Still important, the future of GIS is moving toward its becoming

The Limitless World of GIS Applications

Human creativity knows no bounds. Considering GIS applications, you can quickly find out how true that statement is. The examples in this chapter are clearly just putting a toe into a gigantic ocean of GIS uses that support decisions. The roles of choices and conventions in communicating an understanding of the world are evident in every map.

Below you can find a few more links to additional online sources. I also encourage you as you read this chapter to fire up your imagination and search for GIS uses on topics you are interested in. Reading about these uses and thinking about the choices and conventions people make and follow is a great way to relate what you have read in Chapters 1 through 3 and connect the examples with the principles and practical issues other chapters cover in more detail.

Additional Examples

(Agriculture) *NASA Landsat Satellites Find the "Sweet Spot" for Crops*

(Campaign contributions) *Visualizing presidential campaign contributions throughout Minnesota—MinnPost*

(Environmental protection) *Google Lat Long: The Surui Cultural Map*

(Gerrymandering) *The Great Gerrymander of 2012—NYTimes.com,* February 3, 2013

(Health) Tracking influenza: *www.cdc.gov/flu/weekly/overview.htm#Summary*

(Logistics and disasters) *www.gpsworld.com/logistics-gis-and-disaster-response*

(Retail/sales) *Indoor Cellphone Maps for Mall or Airport—NYTimes.com,* October 11, 2010

(Sea level rise) *NC Considers Making Sea Level Rise Illegal: Plugged In, Scientific American Blog Network*

(Urban disparities) *www.ncbi.nlm.nih.gov/pubmed/19190579*

as integral to spatial enablement in society as maps once were. The many uses of GIS just keep increasing.

Some Examples of Choices and Conventions

The rapid growth of GIS in over some 40 years and the wide use of GIS in applications reflected in a $150 billion or more economic sector worldwide means that any examples, not just these examples here, reflect a very limited range of uses. To round up the introductory material in Chapters 1–3 the examples here have numerous parallels but focus on different choices and conventions that point to fundamental concepts in working with GIS.

Example 1: Tracking Influenza

Every year millions of people become ill with the flu, also known as influenza (see Figure 4.1). In some years, susceptible individuals become grievously ill, some even will die as the result of flu, or related health problems. The availability of GIS has drastically changed the way that specialists and the public understand the spread of influenza each year and the changing health risks for people.

Because influenza is a virus that is constantly changing, it is especially important that health professionals are aware of changes in the strains, can assess the effectiveness of antiviruses, and can identify potential pandemic strains. In the United States, voluntary reporting from over 3,000 sources provides information to address the questions where, when, and what influenza viruses are circulating. A FluView surveillance report appears each Friday at the Centers for Disease Control and Prevention (CDC) website, and interactive systems use maps, charts, and data (see Figure 4.1).

A map of the United States can be viewed showing data at the national, regional, and census-division levels. These regions are choices for the user that reflect conventions that arise in medical care and public health institutions. The pie chart for each area uses categories that also reflect institutional conventions. Finally, a selection bar makes it possible to review data for certain weeks, but not days or months. This choice of a weekly temporal scale may reflect conventions, but quite possibly it is a matter of simplifying the display of this very complex data. Additional data visualizations help specialists gain a deeper understanding when called for.

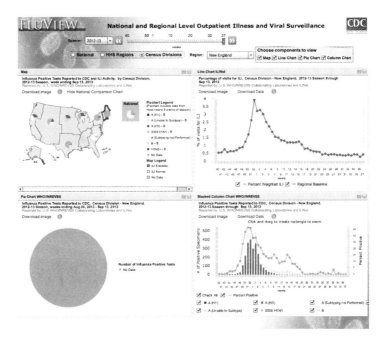

FIGURE 4.1. On the CDC website, the choice of "natural" colors for the background image and the use of red–yellow–green sequence of colors to indicate threats follows established graphic conventions in most of the world for indicating danger levels.

Example 2: Environment Science

In countries around the world, medical specialists have created similar systems to help health care professionals. The environment, threatened or damaged in many areas, needs similar monitoring to enable specialists to plan conservation and respond to threatening situations. Because of the richness and complexity of the environment, many countries, citizens, companies, and nongovernmental organizations (NGOs) have helped monitoring and reporting surveillance that makes key use of GIS. An interesting example is the citizen science monitoring program of the U.S. National Park Service. Monitoring is important to preserve natural resources by recording information about indicator species and observations about trees, land cover and land use, riparian and stream conditions, and many other things. For example, in the Nez Perce park, for seven years high school students from the surrounding area mapped the camas lily, which is associated with seasonal wet prairie ecosystems in the Columbia River plateau of the western United States. The maps they make are valuable for conservation specialists and can be combined with other data (e.g., LiDAR-based topography, to help develop specific management and restoration activities; LiDAR is discussed in Chapter 9).

A 2009 sampling frame map of Weippe prairie shows the location of camas lilies (see Plate 6). The choice was made to show location with black dots, and while it looks like each dot stands for one plant, they indicate sampling areas (quadrates), so the actual number of camas lilies is a prediction based on kriging interpolation (see Chapter 16). This map is an example of straightforward cartographic communication. That north is at the top of the map, that a greenish color is the base map color, that the title of the map is at the top of the page—these are all conventions. Maybe there could be some choice in placement of text or adding explanatory notes, but the clarity of this map comes in no small measure from following conventions. Now, the exact shades of green, the size of the dot symbols, and maybe even the use only of meters in the scale bar could be matters of choice. Fundamentally, those representational matters have a great deal of latitude, but conventions often constrain them to assure clear organizational communication. Other choices, connected to when the data was collected, likely reflect environmental conditions and the camas lilies' growth in a year. Another choice that is not clear from the map alone was the choice of the projection system. Again, given the institutional provenance, it is most likely that the de facto standard projection for National Park service maps was used for this map and others on their website.

Example 3: Travel Information Systems

On highways in the United States and other countries, regardless of whether one is on a long pleasure vacation or a business trip, congestion is something people wish to avoid. In cases where travelers have options, a travel information system can integrate the data from multiple sources in a GIS and then produce a very understandable map or display to help get information for figuring out alternate routes. Many of these GIS are part of more comprehensive information systems that provide information on weather, emergency alerts, and so on. Ease in communication is central here to help people make good and timely decisions. The choices therefore follow conventions: putting north at the top of the map, using red coloring to indicate the most potential danger or slow traffic, using green to show low potential risk of danger or little traffic disruption, showing an almost schematic map of major roads. Using these conventions is a choice, and in dense or large areas the choices may generally be the best possible compromise, but people familiar with the area may benefit from other choices, for example, putting north on the right side of the map to show more detail of an area that is oriented north–south.

Example 4: Small Business Market Analysis

Decisions about how to develop a business involve many factors. While most businesspeople would say that you can't ever get too much information, most of the same people would agree it can be hard to make sense out of the seemingly limitless data sources a businessperson should consider. Many types of analysis are available. Because of the complexity, many also use GIS to clearly communicate the results. For instance, the U.S. Small Business Administration provides an online tool that makes analytical maps to assess your business, map the competition, and find good places to advertise (see Figure 4.2). The maps it produces stay on the conventional side of things in terms of symbology. Even the categories seem to be determined on the software side and given to the map reader. Because of the use of conventions these maps are very plain, and good in communicating a basic message. Considering the complexity, this is a good starting point for business market analysis. Commercial providers offer far more detailed and comprehensive GIS-based solutions with the ability to choose many parameters for market analysis.

Summary

This brief history and a limited consideration of some examples of the many uses of GIS, especially maps produced with GIS, point to an emphasis on problem solving. The importance of choices that align or follow conventions reflects the importance of clear communication. For example, most maps put the north orientation at the top of the screen or paper to ease our reading and understanding of the displayed information. GIS creators of maps generally follow this convention even when another orientation may be better suited to show more detail of an area.

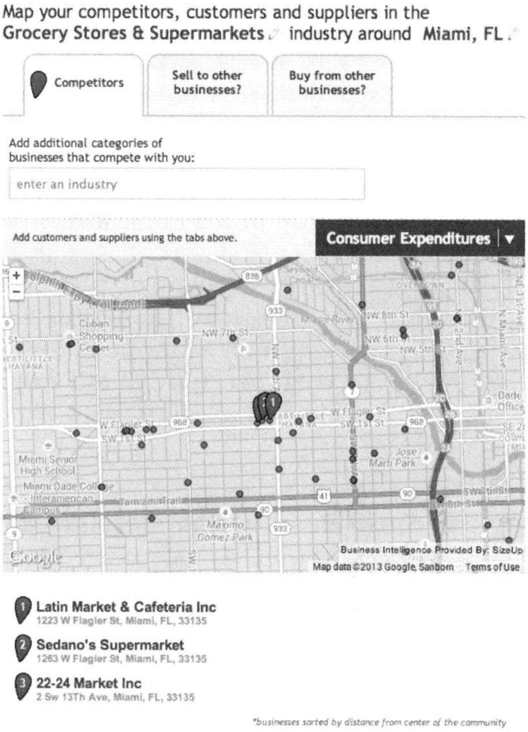

Map your competitors, customers and suppliers in the
Grocery Stores & Supermarkets. industry around Miami, FL.

| Competitors | Sell to other businesses? | Buy from other businesses? |

Add additional categories of
businesses that compete with you:

enter an industry

Add customers and suppliers using the tabs above.

Consumer Expenditures ▼

Business Intelligence Provided By: SizeUp
Map data ©2013 Google, Sanborn   Terms of Use

Latin Market & Cafeteria Inc
1223 W Flagler St, Miami, FL, 33135

Sedano's Supermarket
1263 W Flagler St, Miami, FL, 33135

22-24 Market Inc
2 Sw 13Th Ave, Miami, FL, 33135

businesses sorted by distance from center of the community

FIGURE 4.2. A simple map that follows common conventions for showing areas
forms the background for preparing more complex analytical overviews using this
map-based tool for analyzing markets for small businesses in the United States.

REVIEW QUESTIONS

1. Why do maps remain key media for GI today?

2. Are all maps created with GIS today?

3. What are the advantages of GIS over maps?

4. Why does GIS involve choices and follow conventions?

5. When was the first use of the computer in geography?

6. How large of an economic sector is GIS?

7. Is the global revenue of GIS growing?

8. Why is it important that choices in creating and using GIS usually follow conventions?

9. Are conventions usually simple or complex?

10. Why is problem solving an important emphasis for maps and GIS?

ANSWERS

1. Why do maps remain key media for geographic information today?

   Because of their unparalleled flexibility, accessibility, and allure in any situation, even without reliable electrical power.

2. Are all maps created with GIS today?

   No, but many, maybe even most are.

3. What are the advantages of GIS over maps?

   Greater analytical and representational flexibility; GIS helps solve problems in countless ways.

4. Why does GIS involve choices and follow conventions?

   GIS, as a way of representing and communicating about things and events in the world, involves certain choices and following conventions. Some things are missing, some things simplified, and some things exaggerated to help readers understand what they are seeing and assist GIS users in their activities.

5. When was the first use of the computer in geography?

   In the 1940s.

6. How large of an economic sector is GIS?

   In 2012, the global revenue for the Geo-services sector was between $150 and $170 billion (Oxera).

7. Is the global revenue of GIS growing?

   Yes, and it is expected to continue to grow.

8. Why is it important that choices in creating and using GIS usually follow conventions?

   Conventions help ensure that the representation communicates optimally to the desired audience.

9. Are conventions usually simple or complex?

   As seen in the examples, conventions are usually simple—to give GIS users/readers the best chance of understanding the information presented.

10. Why is problem solving an important emphasis for maps and GIS?

    Helping to solve real-world problems is behind much financial investment in maps and GIS.

Chapter Readings

Chrisman, N. (2006). *Charting the unknown: How computer mapping at Harvard became GIS*. Redlands, CA: ESRI Press.

Dangermond, J. (2008). GIS—Geography in action. *ArcNews, 30*(4), 6–8.

Pick, J. B. (2008). *Geo-business: GIS in the digital organization*. New York: Wiley.

Tomlinson, R. F. (2011). *Thinking about GIS*. Redlands, CA: ESRI Press.

Web Resources

⟳ A recent overview of GIS-related industry developments: The Boston Consulting Group. (2012). Putting the U.S. geospatial services industry on the map. Retrieved from *www.valueoftheweb.com*.

⟳ Providing a long-overdue assessment of economic impacts: Oxera Consulting. (2013). What is the economic impact of geo services? *www.valueoftheweb.com*.

⟳ Also a good overview of GIS industry-related developments: U.S. Department of Labor. (2010). High growth industry profile—Geospatial technology. Retrieved 11 May, 2011, *www.doleta.gov/BRG/Indprof/geospatial_profile.cfm*.

EXERCISES

1. Give Your Own GIS Example

Find a different example from the ones presented of how GIS can be used to solve a problem. Explain how GIS is used. Look for examples that clearly show GIS in use. You might also use some of the web resources in this and other chapters to get started.

2. Choices and Conventions

What choices and conventions are important in the example you found for Exercise 1?

PART 2

Fundamentals and Functions

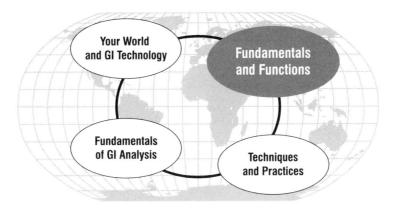

There are several standard sources for the (geographic) information that feeds into a geographic information system, and that data are in predictable forms, such as locational coordinates. The second part of this book, Fundamentals and Functions, covers those key sources and forms that data take, many of which carry over from the practices of cartography, map-making. Coordinate systems, the storage of data in databases, cartography, methods of data acquisition, data acquired from remote sensing, and how data is/are represented visually will usually be present when producing maps and other "products" within the framework of a GIS.

These chapters by themselves are meant to serve as a standalone introduction to the building block components of GIS. If you have studied some or all of this material before, you may be using this section of the book to refresh your knowledge or for a workshop. If you have worked through the chapters of Part 1, these chapters of Part 2 start to fill in the detail of an understanding of geographic information systems. By their nature, GIS always draw from a multifaceted body of knowledge and use a varied group of "tools."

The components that make up a GIS, as covered here in Part 2, eventually make up a GIS when it is up and running. And the issues of functioning GIS, and what can be done with them (analysis), are the subjects of Parts 3 and 4.

CHAPTER 5

Projections

This chapter focuses on the principles connected to projections—the transformation of spherical coordinates to planar coordinates—you will need for work with GI for making maps and other purposes. Chapter 4 also presents some historical background and specific details of various projections.

Projections occupy one of the most essential roles in cartography for geography and GI. For some people, this role may arguably be perhaps the most essential, because most GI is "projected," even if the projection information never shows up on a map. This has started to change as more and more GI is collected and stored in latitude and longitude coordinates, which are not projected and commonly used in online mapping. But even if all the data you need and want is available in latitude and longitude coordinates, you will probably need to project it to make the sort of map that people are familiar with, or combine it with data collected using another projection.

Maps without Projections

Some people would claim that if a thing or event is shown on a map, it must be projected. In most cases this is true—and for good reasons. But there are exceptions. These exceptions are important enough to pay attention to. The first exception was already mentioned: locations stored in latitude and longitude coordinates are not projected—they are spherical coordinates. It's even possible to make a planar (flat) map with these coordinates in a grid, but such a map is greatly distorted and as a result can be misleading. The second exception is the maps drawn following artistic or design criteria rather than scientific concerns. Usually these maps are used for advertisements, but they can also be used to show transportation networks, to illustrate tourist destinations, and to serve other popular forms of communication. The third exception, globes, is a nonprojected way of showing things, events, and relationships without the distortion of projections. A global **tesselation** using hexagons or octahedrons to subdivide the sphere is another nonprojected way of representing a round surface.

A Brief History of Projections

The reasons for using projections go back to desires to accurately represent the spherical surface of the earth on flat maps. For GI and a map to be useful, the locations and relationships must be accurate. The uses of a nonprojected map using latitude and longitude coordinates (historically these were determined by the use of sextants, cross-staff, etc.; today they're mostly determined by GPS—see Figure 5.1), or an advertising map showing simple directions, are limited by their inaccuracy. You can use such a map with directions to the new amusement park to find your way there, even if you're not from the area, but you can't use it in most cases to navigate to the beach or swimming pool. Its limits mean you won't be able to discover and understand the relationship of the amusement park to things and events not shown on that map. Importantly, because of the curved surface of the earth, nonprojected maps of larger areas showing locations and sizes of things and events would be inaccurate. The Euclidean geometrical measure of the distances on the earth's surface, which is the most common geometry—already practiced by the ancient Egyptians—cannot take its curvature into account, but instead uses a Cartesian coordinate system and a projection that have already taken the spherical shape of the earth into account. Thinking about other possible uses, projections make it easier to compare GI and maps of the same area because they provide a framework for people and organizations to systematically locate things and events.

As you can probably already imagine, it is no surprise that the first maps were based on work by geographers who were locating things and figuring

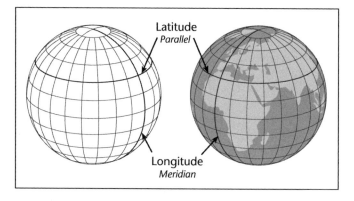

FIGURE 5.1. Latitude and longitude lines.

out relationships among events. Ptolemy (c. 100–168) wrote the book *Geography* with the location of cities, coasts, and other important places of the world known to the ancient Greeks. The Romans may have used this book for making a map that showed, in a greatly distorted manner, Europe, North Africa, the Near East, and India, even indicating China. The original and all copies of this map were lost, except for one, a re-creation done in the 15th century that is now known as the Tabula Peutingeriana.

While maps such as the Tabula Peutingeriana and many others had been used for a very long time, maps that show location accurately have only been around for some 400 years, once it was discovered how to determine longitude. Cartographers until then could only accurately determine the latitude of places. This means that while the equator could commonsensically be calculated as the halfway place between the north and south poles, the 0° starting measure for longitude was only agreed to in the late 19th century and placed in Greenwich, England. Up until then, 0° longitudes started from different locations including Paris and the Faro Islands. Knowing where the starting measure of longitude is located is crucial for accurate navigation. Before there was widespread agreement about where the starting (0°) longitude is for all people and countries, an arbitrary starting longitude was fine so long as it was used systematically.

Roles of Projections

One of the key roles of projections has been in the production of maps for navigation, naval or aeronautical in most cases, which are called **charts**. The development of accurate ways to determine location went hand-in-hand with the growth of European naval powers. However, because these are spherical coordinates, and mariners needed flat maps to take with them, projections became crucial. The Mercator projection is perhaps so commonplace because a straight line in this projection shows a constant compass bearing. You should remember that there are many other projections, but the Mercator projection possesses the quality that lines of a constant direction are straight lines.

Because of this character, the Mercator projection was very important for navigation on water by compass, but other modes of transportation can better use other projections. More recently, since airplanes began to fly regularly across and between continents, another type of projection was needed for their navigation. A line of a constant compass direction may be straight in the Mercator projection, but this line does not show the shortest distance. The shortest route for an airplane high above the earth's surface is not a straight line, but a line on a sphere, called the **great circle distance** (see Figure 5.2).

Different projections are used for maps with different roles. The size of the area to be mapped, the desired **projection properties**, and the characteristics of the GI and map are the key determinants. The size of the area

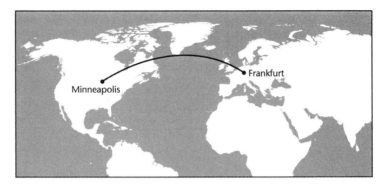

FIGURE 5.2. Great circle path between Minneapolis, USA, and Frankfurt, Germany. The great circle distance is 4,392 miles.

distinguishes basically between the whole world, a continent, a state or province, a region, a county or city, and still smaller units. Different projections fit different areas better or worse, depending on their use. The Mercator projection is quite inaccurate for comparing the sizes of areas because of distortion near the poles, but is quite useful for maps used in navigation. A projection property refers to whether the projection represents angles, areas, or distances (from one or two points) as they are found on the surface. No projection retains both angles and areas. A projection can retain one projection property—for example, the Mercator projection preserves angles. All **transverse** (turned 90° to be oriented north–south) Mercator-projected GI and maps are useful for mapping north–south oriented small areas because this projection is conformal and also preserves shapes over small areas along the line of tangency where the projection theoretically touches the earth's surface. The characteristic of the GI or map indicates how the projection should show geographic relationships and scales. Choosing a projection that preserves one projection property often leads to other **distortions**. For an individual state or province, a projection that maintains constant area to make comparisons of areas possible is beneficial, even if some shapes over a larger area may begin to look distorted. Indeed, larger areas are hard to show without distortion in any case; many projections commonly used for world maps compromise and distort both area and shapes. Why distortion is commonplace for projections, what are the projection properties and characteristics, and how to choose a projection is discussed later in this chapter.

Making Projections

Even if you are only going to use maps and will never work with GI, you need to know some important things about projections. The first is that projections make use of different models of the earth. Generally, projections for the entire earth use a simple spheroid. When dealing with maps or GI of the entire world, the loss of accuracy is slight compared to the resolution of the

Making Projections with Light

Although most projections are calculated mathematically, the underlying transformation from a three-dimensional to a two-dimensional representation of all projections can be physically constructed with the aid of a few common items: a light (flashlight or lamp), a two-liter plastic bottle, a lampshade, and a piece of wax paper or flat plastic you can draw on. You will write on all of these items, so you need to be sure they are no longer needed.

To make the construction surfaces, you will need to prepare the plastic bottle by cutting off the top and bottom carefully with a scissors or knife. The lampshade and the flat wax paper or plastic are ready to be used as they are. On each of these objects you should mark a series of horizontal and vertical lines. On the lampshade and piece of wax paper or plastic, they should radiate from the center. On the lampshade, they should, if extended, meet each other at an imaginary point above

the top of the lampshade; on the wax paper or plastic, they should radiate from a circle located at the center.

The construction surfaces you made correspond to the developable surfaces used in cylindrical, conic, and planar types of projections. To show how each developable surface is used, take a flashlight or light placed at the middle of the bottle or lampshade or behind the wax paper or plastic surface and shine the light source at a nearby wall or piece of paper. (It usually helps to dim the room lights when you do this.)

What you see on the wall or paper is the projected surface that corresponds to each type of projection. Try moving the light, the paper, and the construction surface to see how the changes affect each projection. These changes correspond to parameters used in the construction of map projections discussed in this chapter.

GI or detail of the map. Projections needed for more detailed purposes or smaller areas of the earth use an *ellipsoid* (see Figure 5.3), with even more axes, that generally fits the actual shape of the earth. For very detailed purposes and the highest levels of accuracy, people use a *geoid,* often optimized for the shape of the earth in one particular and relatively small area (see Figure 5.4).

While this may seem needlessly complex, you should remember that because the earth is constantly changing shape (and not only from volcanoes

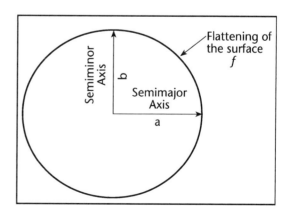

FIGURE 5.3. Reference ellipsoid showing major parameters.

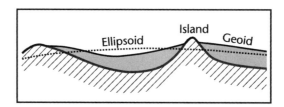

FIGURE 5.4. In this schematic drawing, an ellipsoid and a geoid represent the earth's surface. The ellipsoid is less accurate than the geoid, but both may not properly align with actual locations distant from their geospatial optimizations.

and earthquakes, but gradual movements that are imperceptible to people), different uses need different levels of positional accuracy. On one extreme, a map showing worldwide the most visited tourist sites for the last 10 years needs very little accuracy; on the other extreme, an engineer's plan of a 2-km tunnel for a new railroad needs extremely great accuracy. Most GI and mapping activities need a level of accuracy somewhere in between—often the cost of preparing the GI and the available budget determine the accuracy.

What makes all this complicated for working with GI and maps is that there is no standard earth model, or geoid model, or spheroid model, or ellipsoid used to represent locations on earth. The use of different models makes it paramount for GI users to know the model used for projecting the GI, which is often called a "datum" (see below for more information about datums). Work on specific models of the earth is done by geographers known as geodesists, and information about geoids and datums is geodetic information.

The Geoid Model

The most accurate model of the earth's surface is the geoid. The earth, because of its constantly changing shape due to tectonic movements and undulations of its gravity field, can be described in the most detailed fashion through sets of measurements that are used to produce a geoid. The expert geodesists who determine geoids and their constants put the geoid model into relationship with the planetary body or extremely detailed information about elevations in a particular area. Geodesists describe a *geoid* as the equipotential surface of the earth, which means the known earth's surface under consideration of different local strengths of gravity resulting from different masses of the earth's geological makeup, fluctuations in the earth's core, and other factors. For example, the Marianna Trench in the Pacific Ocean and the large bodies of iron ore found in parts of South Asia, Sweden, and many other places both locally affect the shape of the earth's surface because of the lessened or increased pull of gravity due to the lesser or greater mass at those locations. Basically, what geodesists consider is how differences in the earth's gravity affect the shape and size of the earth. For instance, denser material in the earth's crust, such as iron, influences gravity more than lighter sedimentary rocks do. The geoid takes these and other differences into account. These differences are measured in millionths of the earth's normal gravity, which seems small, but the effects on the shape of the earth and location measurements can be large. You can think of the geoid as a collection of many gravity vectors, individual gravity forces, each of which is perpendicular to the pull of gravity, that distort the earth's shape from the ideal, yet imaginary, sphere.

Practically, the geoid was until recently only used for specialized purposes for relatively small areas. Geoids were almost always calculated for smaller areas because of the complexity and cost of collecting the necessary data. With the advent of satellites and improved measuring devices, however, data collection has become much easier and geoids have become more common. They are the reference standard when working with global positioning systems (see Chapter 7). The geoid provides vertical location control. Geoid positions usually refer to a reference ellipsoid for horizontal location control and vertical location control. Differences between the ellipsoid positions and the geoid positions are called "geoid undulations," "geoid heights," or "geoid separations." The horizontal and vertical locations of the projection surface based on an ellipsoid can be adjusted to the irregular shape of the geoid compared to the regular mathematical surface of an ellipsoid through geodetic techniques.

The Ellipsoid Model

The ellipsoid (also called a spheroid in some cases) is the most commonly used model for projections of GI and maps. It includes the noticeable distortion between the length of the earth's north–south axis and its equator, which bulges a small amount due to the centrifugal force of the earth's rotation. In the simplest mathematical form, it consists of three parameters:

- An equatorial semimajor axis a
- A polar semiminor axis b
- The flattening f

Mapmakers and geodesists have produced many ellipsoids. John Snyder wrote that between 1799 and 1951 twenty-six ellipsoid determinations of the earth's size were made. Each of these ellipsoids has a history and sheds light onto the science, culture, politics, and personalities involved in establishing the ellipsoid through complicated and challenging field survey coupled with exhaustive calculations. Ellipsoids were developed to have a more accurate reference for mapping, to satisfy individual ambition, to serve national goals, to make more accurate measurements, and so on. The surveys conducted to create ellipsoids were often ambitious expeditions into the remote areas of the world and continue to provide the material for many stories. Multiple ellipsoids were developed and refined as measurements improved, and ellipsoids have often been specially defined for specific areas—for example, for U.S. counties. Working with data or maps from different periods often involves determining if different ellipsoids were used in collecting data; data from different coordinate systems, even if in the same area, may also have different ellipsoids (see Table 5.1).

TABLE 5.1. Selected Ellipsoids Parameters

Name	Semimajor axis	Semiminor axis	Flattening
Bessel (1841)	6,377,483.865 m	6,356,079.0 m	1/299.1528128
Clarke (1866)	6,378,206 m	6,356,584 m	1/294.98
Krassovsky (1940)	6,378,245 m	6,356,863.03 m	1/298.3
Australian (1960)	6,378,160 m	6,356,774.7 m	1/298.25
WGS (1984)	6,378,137 m	6,356,752.31425 m	1/298.257223563

The Spheroid Model

The *spheroid* is the simplest model of the earth's surface, using only a single measurement to approximate the shape of the earth's surface for GI and maps. This measurement is the distance from the hypothetical center of the earth to the surface, or, in geometrical terms, the radius. The mean earth radius RE is 3,959 miles (6,371.3 km). It is very inaccurate for many GI uses and you should only use the spheroid for scales smaller than 1:5,000,000,000. The inaccuracies of this model of the earth's surface are not apparent at these small scales. It is much easier to calculate the projections using a spherical model but using the spheroid for projecting GI and making maps for scales larger can lead to grave inaccuracies.

Putting the Models Together: Demythologizing the Datum

Datum is the term used to refer to the calibration of location measurements including the vertical references, horizontal references, and particular projections or versions of projections—for example, the North American Datum 1927 (NAD 27) or the North American Datum 1983 (NAD 1983). Datums constitute one of the most confounding aspects of working with projections for many GI and map users. For most intents, this term simply specifies the model of the shape of the earth at a particular point in time and often for a particular area—for example, North America, Europe, or Australia. A horizontal datum is often the basis for determining an ellipsoid used in a projection for a coordinate system (see Chapter 6). A datum can be used with different projections—for example, the NAD 1927 is used with both the Lambert and the transverse Mercator projections. For GI users, datums are references to a set of parameters needed for measuring locations and the basis for projections. Because there are many parameters and the mathematics for transforming datums is highly complex, many people have been stymied by datums. But it is really, for most general purposes, quite simple: the datum refers to a reference surface for making positional measurements. While most datums in North America are described in technical guidelines or even laws, theoretically a datum can be defined by any government agency or private group as it sees fit.

Datums distinguish between horizontal and vertical references and local and geocentric datums. A datum should (but might not) contain both horizontal and vertical references. Horizontal references are used to measure the location of positions on the earth and vertical datums are used to measure the elevation of a position. You can think of a vertical datum as the base level used in recording elevations or the mean height of tides. All elevations using the vertical datum are related to this zero elevation. Local datums, in fact, are used for areas up to the size of continents—for example, the NAD of 1927, which made a location on Meades Ranch in Kansas the starting point of the triangulation that measured the earth's undulations and put them into relationship with the Clarke 1866 ellipsoid. Geocentric datums—for example, the World Geodetic System Datum of 1984—take the entire earth into consideration and lack an origin point; they don't have a defined datum point, but are calculated from a network of geodetic observations. The difference between local datums can be several hundred meters—for instance, between NAD 1927 and NAD 1983 in some areas of the United States. Conversions of measurements between the two systems can become quite complex. Fortunately, programs are widely available to transform between popular datums—for example, between NAD 1927 and NAD 1983—for most areas. Datums are constantly being changed and updated. Currently the most up-to-date U.S. horizontal datum is the North American Datum of 1983 (NAD 1983) after the U.S. National Geodetic Survey's (NGS) National Adjustment of 2011 and the most up-to-date U.S. vertical datum is the North American Vertical Datum of 1988 (NAVD 1988). The NGS has further plans to replace these two datums with new geometric and vertical datums including changes in position with time, adding a fourth dimension to the system. A few important datums in North America and globally are listed in Table 5.2.

Types of Projection and Their Characteristics

Theoretically, the number of possible projections to transform coordinates from a spheroid, ellipsoid, or geoid to a planar coordinate system or flat map is unlimited; practically, the number is limited only by the creativity of

TABLE 5.2. Selected Datums

Horizontal datum name	Ellipsoid	Local/ geocentric	Where used
NAD 1927	Clarke 1866	Local	North America
NAD 1983	GRS 1980	Geocentric	North America
WGS 1984	GRS 1980 with additional measurements	Geocentric	World
New Zealand Geodetic Datum (NZGD) 2000	GRS 1980	Geocentric	New Zealand

mathematicians and geodesists and the needs of organizations to coordinate their creation, maintenance, and use of GI for the many public and private uses. To start understanding projections, you should familiarize yourself with the three basic developable surfaces, also called "projection families," used to create map projections. **Developable surfaces**, which are an actual or imaginary drawing of the projection, were used to help cartographers visualize the projection process (see Figure 5.5). They are no longer used to project maps, but they are helpful in understanding projections.

Developable surfaces can be drawn, but many projections are created without them. Projections created with developable surfaces can be demonstrated using a light hung in the middle of a transparent globe or by shining a flashlight through a portion of a globe onto the developable surface. For example, a two-liter plastic bottle, cut off at both ends and marked with a constant interval of vertical lines, with a light bulb hung in the middle to project the lines on a wall, will show how a cylindrical projection projects latitude and longitude on a flat surface. All projections using pseudo-developable surfaces can only be described mathematically. They cannot be created in any mechanical manner.

A key characteristic of all projections, whether developable surface or pseudo-developable, is called aspect (see Figure 5.6). The **projection aspect**

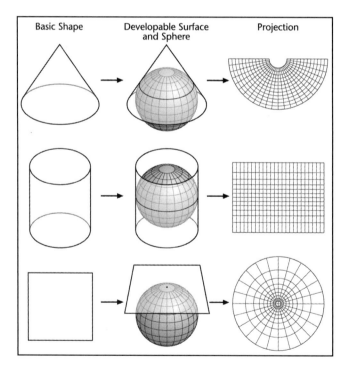

FIGURE 5.5. Basic geometric shapes (cone, cylinder, and plane) serve as developable surfaces, shown here with a reference globe. The resulting projections of latitude and longitude lines are shown in the rightmost column.

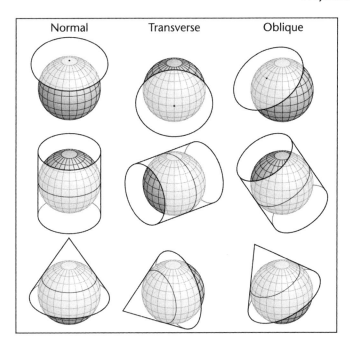

| Normal | Transverse | Oblique |

■ FIGURE 5.6. Some possible aspects for conical, cylindrical, and planar projections.

refers to the orientation of the developable surface to the earth. Various conventions have come and gone in cartography over time. For future users of GI, I think it is most pragmatic to distinguish among equatorial, transverse, oblique, and polar aspects. The differences refer either to the orientation of the projection to a region of the earth (equatorial or polar) or to the developable surface of that type of projection—for example, transverse Mercator projections are rotated 90° from the Mercator projection's usual equatorial orientation. The basic differences are best visualized in a figure showing the different aspect for each developable surface (see Figure 5.6). The consequences for distortion and accuracy are discussed later in this chapter.

Some possible aspects for conical, cylindrical, and planar projection include equatorial and polar. Equatorial orientation has the projection's center positioned somewhere along the equator. Polar aspect occurs only with planar projections. All three projections may have an oblique aspect (based on Jones, 1997, p. 75).

Tangent/Secant

Figure 5.7 illustrates differences in how projections "touch" the developable surface of a reference globe, another important characteristic of projections. These places of contact between the developable surface and spheroid, ellipsoid, or geoid are the most accurate for any projection and are called *standard parallels* or *standard lines*. Tangent projections "touch" the reference

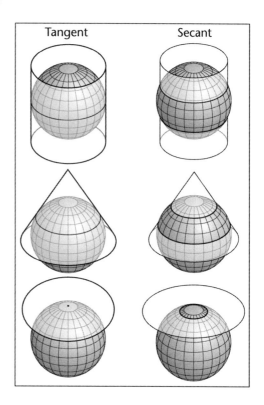

Tangent   Secant

FIGURE 5.7. Examples of tangent and secant projection surfaces.

globe at one point or along one line. Secant projections "touch" the reference globe along two lines or in an area.

Projection Properties

Projections alter the four spatial relationships (angles, areas, distances, and direction) found on a three-dimensional object. As mentioned earlier in the overview of projections, most projections only maintain one of the properties in a specific manner—for example, equidistant projections preserve distance from *one* point to all other points. Many projections, especially projections used for larger areas, compromise all these properties.

The projections that preserve angular relationships from one point are called *conformal*, but you should remember that conformal refers to the preservation of angles only, never shapes. Figure 5.8 includes a Lambert conformal conic projection, which preserves angles, but not areas. If a projection preserves areas in the projection by a constant scaling factor, it is called an *equivalent* projection. Equivalent projections preserve areas, but not shapes. The shapes of continents or countries can change in an equivalent projection, but their areas correspond to the actual areas on the earth (Figure 5.8, Sinusoidal projection). Projections that preserve distances from one or two points to other points are called *equidistant* (Figure 5.8, Stereographic projection). The projections that preserve directions are called *azimuthal*, or true direction projections. Directions are only preserved from the center of the map in azimuthal projections.

Projections that are neither conformal nor equivalent are called *compromise* projections. They are usually developed to make more graphically pleasing maps and do this by finding a balance between areal and angular distortion (Figure 5.8, Robinson projection).

Some Common Projections, Characteristics, and Uses

With so many projections, it is certainly possible to find a projection for every occasion. Fortunately, for most GI uses, the projections are already determined. The choices for maps, especially maps of large areas, are much broader. The following examples highlight a few widely used projections for each of the four projection properties.

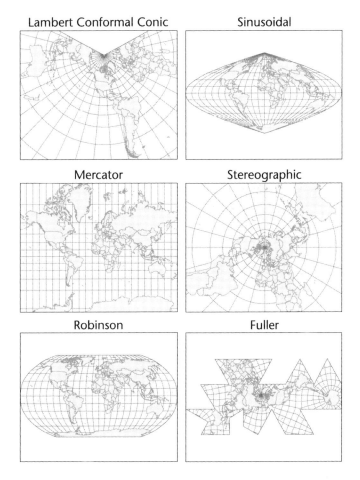

FIGURE 5.8. These six different projections show the countries of the world.

Lambert   The Lambert conformal conic projection preserves only angles. Used for mapping continents or similar areas, it is commonly used for areas with an east–west orientation—for example, the continental United States.

Sinusoidal   The sinusoidal equal area projection preserves areas, but distorts angles and shapes. It is used for maps showing distribution patterns.

Mercator   The very common Mercator projection is a conformal projection with the very unusual quality of showing lines of constant bearing (called *loxodromes* or *rhumb lines*) as straight lines. This made the Mercator projection very valuable for sailors, who could use one single compass heading to determine the direct route between two points. Transverse Mercator projections are widely used for areas with north–south primary orientations (see Figure 5.9).

Stereographic   The widely used stereographic projection is an azimuthal projection developed in the 2nd century B.C.E. that preserves directions; it is a further development of much older stereographic projections. It additionally has the particular quality of showing all great circle routes as straight lines; however, directions are true from only one point on the projection. It is used usually to show airplane navigation routes.

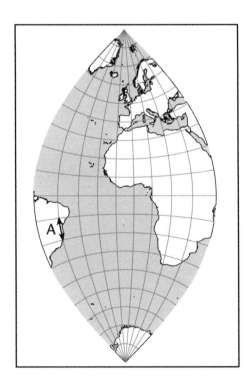

FIGURE 5.9. A transverse Mercator projection.

Robinson   The Robinson projection is a compromise projection that
                 fails to preserve any projection properties. It is graphically
                 attractive; it was adopted by *National Geographic* in 1988 and
                 is widely used elsewhere.

Fuller   The Fuller projection was introduced in 1954 by
                 Buckminster Fuller. It transforms spherical latitude and
                 longitude coordinates to a 20-sided figure called the
                 icosahedron.

Calculating Projections

Examining the mathematics of projections is helpful for grasping how
a projection transforms locations measured in three dimensions to two-
dimensional locations. You should always note that projections are never
transformations between two two-dimensional coordinate systems, but
between locations found on or near the surface of the three-dimensional
planet earth to a two-dimensional coordinate system.

The three examples examined here are widely used. The sinusoidal
projection is a pseudo-cylindrical projection developed in the 16th century;
the Lambert conformal conic projection is widely used around the world
for east-to-west-oriented areas; the Mercator projection is very common.
However, the mathematics for each map projection discussed here are quite
straightforward, especially since these examples are based on spheroids.

Sinusoidal Projection

The sinusoidal projection is a simple construction that shows areas correctly,
but shapes are increasingly distorted away from the central meridian. Parallels of latitude are straight, and longitudinal meridians appear as sine or
cosine curves.

The equations for calculating the sinusoidal projection are quite simple.
You only need to remember to use radians for the angle measures of longitude and latitude and to place a negative sign in front of longitude values
from the western hemisphere.

Equations for calculating a sinusoidal projection:

$$x = R\lambda(\cos \phi)$$
$$y = R\phi$$

where ϕ is the latitude, λ is the longitude, and R is the radius of the earth
measured at the scale of map.

Lambert Projection

The cylindrical equal-area projection shown here is one of several projections that Lambert developed in the 18th century. It remains a widely used

projection, especially in atlases showing comparisons between different countries or regions of the world. Polar areas have strongly distorted shapes, but most continents evidence only minor distortion.

Below are equations for calculating a Lambert cylindrical equal-area projection:

$$x = R\lambda$$
$$y = R \sin \phi$$

where ϕ is the latitude, λ is the longitude, and R is the radius of the earth measured at the scale of map.

IN DEPTH **Calculating Projections with Radians**

You may need to use radians for an exercise calculating projections or for other angular measures. The sinusoidal projection, many other projections, and other measures involving angles are often calculated with radians, which is another form of angle measures: $1° = \pi/180$ radians, $360° = 2\pi$ radians. Radians indicate the length of that part of the circle cut off by the angle, and make it easy to determine distances on circular edges or round surfaces.

$$\text{radians} = (\text{degrees} \cdot \pi)/180$$

The length of part of a circle (called an arc) is determined by multiplying the number of radians by the radius. For example, the length of an arc defined by an angle of $10°$ on a circle with a 100-m radius is 0.1745.

1. Determine radian measure of angle:
$$n \text{ radians} = (10° \times \pi/180)$$
$$n \text{ radians} = 0.1745$$

2. Calculate length of the arc:
$$\text{arc length} = n \text{ radians} \times \text{radius}$$
$$\text{arc length} = 0.1745 \times 100 \text{ m}$$
$$\text{arc length} = 17.45$$

Some common angle measures in degrees and their equivalents in radians are listed here.

Degrees	Radians
90°	$\pi/2$
60°	$\pi/3$
45°	$\pi/4$
30°	$\pi/6$

Mercator Projection

The very common Mercator projection uses only slightly more complicated equations.

Below are equations for a Mercator projection (Snyder, 1993):

$$x = R\lambda$$
$$y = R \ln \tan (\pi/4 + \phi/2)$$

where λ is the longitude (– if in the western hemisphere) for determining values of the y axis and ϕ is the latitude (+ if north, – if south of the equator) and R is the radius of the earth measured at the scale of the map. The term *ln* refers to the natural logarithm to the base e. All angles again are measured in radians.

Distortions

Distortions arising through projections are unavoidable. They have significant consequences for accuracy, so it helps to know more about distortions in order to choose the best projection for different purposes and to be able to take distortions into account.

As a general place to start out, we can categorize distortions in terms we have already seen: the four projection properties of angles, areas, distances, and direction. Many projections distort one or two of these projection properties. Distortion of angles (including shapes) is sometimes easy to detect, especially for large areas when familiar shapes of states, continents, or even provinces are distorted; but projections of small areas may lack readily visible evidence of distortions and require the use of special graphics or statistical measures to determine the distortions. The same applies to areas. The distortions arising related to distance can be significant because, as you know now, no projection for large areas accurately shows distances for all points, but can only be accurate for a few points. Small areas are another matter, but you still should check to see what distortion a projection creates. Direction can likewise be distorted in a subtle fashion that is not visually noticeable, but is of significance should the map be used for navigation purposes.

One easily overlooked source of distortions is the difference between the datums, geoids, and ellipsoids used in creating different GI or maps. Even if GI or a map is made by the same agency or company using the same projection, a change in the datum, geoid, or ellipsoid can lead to distortions when compared with other GI or maps for the same area.

Describing Distortions

To describe and assess distortions, it is useful to determine the scale factor (SF) at different places on a map. By comparing scale factors with map scale

The Case of Changing Projections and Datums in Belgium

Not only famous for beer and chocolate, and as the site of much of the European Union administration, Belgium has a rich history in geography and cartography. Datums specify the earth's size and shape as well as specifying the origin and orientation of a coordinate system. This brief history of changes made to projections and datums in Belgium shows why datums change. When two geographic areas (countries) use different projections and datums, points on the border between the two areas will not match up.

After World War II Belgium adopted a new geodetic system, which was based on the Hayford ellipsoid (1924), with its initial point located at the Brussels Observatory. The Lambert conformal conic projection with two standard parallels was the projection chosen to create Belgium's cartographic grid. On this basis a cartographic series at the scale of 1:25,000 was established. By 1972, it had become necessary to create a new geodetic datum, Belgium Datum 1972, which was based on a new global compensation. This datum also used the Hayford ellipsoid, but the initial point had shifted since 1950. New project parameters were defined, constituting the Belgian Lambert 72 (BD72) system. Most current topographical maps in Belgium use the BD72 coordinates. The system can only be used in Belgium.

Spatial geodesy's advances in the second half of the 20th century have made it possible to determine and track over time the shape of the geoid and position of the earth's center of mass with great accuracy. (They are constantly changing, notably due to plate tectonics.) The center of mass is the starting point for a system with three perpendicular axes (x, y, z). Two are on the plane of equator and the third corresponds to the direction of the poles. A point may be localized by any triplet of Cartesian coordinates and may be below, above, or on the surface of the earth. The International Terrestrial Reference System (ITRS) and World Geodetic System (WGS) are based on this principle. A reference framework based on satellite instruments and earth observations determines the x, y, z coordinates of several hundred points on earth. The resulting "Frame" is referred to by the acronym (initialism) of the reference system (e.g., ITRS) followed by the year of the observations. In Belgium and other countries, geodesists globally adjust ellipsoids to the shape of the geoid and center them on one center of mass, which guides how longitude and latitude are determined. The Geodetic Reference System (GRS80) of 1980 was followed by the World Geodetic System of 1984 (WGS84); it had a slightly different flattening coefficient and definition of the geodetic datum and ellipsoid. The problem of different countries having different geodetic systems resulted in a push to standardize the use of WGS84. More recently, Belgium has begun to implement European Terrestrial Reference System 1989 (ETRS89). This is the EUREF system recommended for European cartographical and topographical activities. In Belgium a GPS collection of point locations for around 4,200 points using ETRS89 coordinates was used to create the Belgian geodetic frame, or BEREF. The GRS80 ellipsoid is associated with BEREF and was used as the basis for the new 2008 Belgian Lambert projection. Once the BEREF was completed, a new version of Belgian Lambert went into use with the coordinates of the central meridian and standard parallels (49-50 N and 51-10 N) defined on the global ellipsoid GRS80 and lined with ETRS89. To avoid confusion with the 1972 version, the coordinates were given a false origin in the 2008 version by adding approximately 500 km. Though projections and datums must change due to changes in the earth's crust, discretionary decisions can be made, to help avoid confusion among the different systems in use.

Source: National Committee of Geography of Belgium. (2012). *A concise geography of Belgium.* Ghent, Belgium: Academia Press.

at the standard point or standard lines you can assess the scale distortions using this formula:

$$\text{Scale factor} = \frac{\text{Local scale}}{\text{Principle scale}}$$

where *local scale* is the scale calculated at a particular place and *principle scale* is the scale computed at the standard point or a standard line (see Table 5.3).

For example, the scale factor of a transverse Mercator projection with a principle scale of 1:400,000,000 calculated between 20° and 30° S will indicate how much distortion the projection introduces. First, calculate the local scale by measuring along the meridian between 20° and 30° S. This gives you the map distance, which is 3.1 cm (1.2 in.) (shown in Figure 5.9 with the letter A). You compute the ground distance between the same portion of the meridian by consulting a table showing the lengths of a degree of latitude along a meridian. At 20° a degree of latitude is 110,704.278 m long. Multiplying the 10° of latitude to 30° S would measure approximately 1,107,042.78 m or 1,107.04 km. Second, by substituting the 3.1 cm and 1,107.04 km into the map scale equation, you can calculate the local scale:

Map scale = earth distance/map distance
Map scale = 1,107.04 km/3.1 cm

The units in the equation must be equal, so you first need to convert kilometers to centimeters by multiplying by 100,000.

1,107.04 km × 100,000 = 110,704,000 cm

Calculate the local map scale:

Map scale = 110,704,000 cm/3.1 cm
Map scale = 1:35,710,967.74

TABLE 5.3. Table of Meridian Distances for Various Latitudes

Latitude (°)	Miles	Kilometers
0	68.71	110.57
10	68.73	110.61
20	68.79	110.70
30	68.88	110.85
40	68.99	111.04
50	69.12	111.23
60	69.23	111.41
70	69.32	111.56
80	69.38	111.66
90	69.40	111.69

This map scale is considerably larger than the map scale along the standard line of 1:30,000,000. You can now compute the scale factor using the scale factor equation:

$$\text{Scale factor} = \frac{35{,}710{,}967.74}{30{,}000{,}000}$$

$$\text{Scale factor} = 1.19$$

This scale factor suggests that the distances in the transverse Mercator projection increase away from the central meridian. A visual check of the projected map supports this conclusion.

Tissot Indicatrix

A visual analytic to display and examine projection distortions was developed by the mathematician Nicholas Tissot in the 19th century. The concept is simply that any small circle on a spheroid or ellipsoid, when projected to the same point on the flat map, will show the distortion created by the map projection for that area through the projected shape and size of the circle. When the circles are plotted at various points on a map, they allow for a visual comparison of distortion. You should note that the changed shapes and sizes of the indicatrix refer to individual points and cannot be used in evaluating distortion of continents or water bodies.

The indicatrix has two characteristics that can be used to evaluate distortion. The first is the two radii, semimajor (a) and semiminor (b), which are perpendicular to each other (see Figure 5.10). The semimajor axis is aligned in the direction of the maximum SF and the semiminor axis is aligned in the direction of the minimum SF. The second is the angle between two lines l and m that intersect the center of the indicatrix circle, but are turned 45° in respect to the center, if there is no angular distortion. The distances of the semimajor and the semiminor axes, respectfully, indicate the scale factor distortion along each axis. The angle between two lines l and m indicates the amount of angular distortion. For example, a circle where l and m intersect

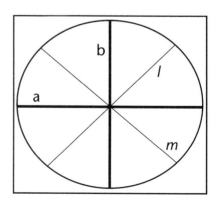

FIGURE 5.10. Tissot's indicatrix circle indicating no areal and no angular distortion.

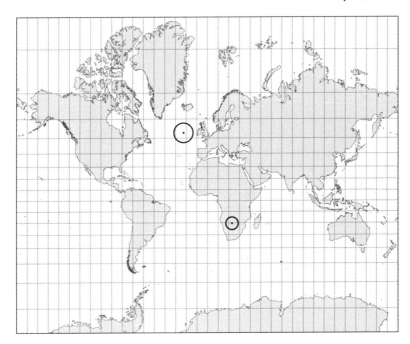

FIGURE 5.11. Two Tissot indicatrix circles shown on a Mercator projection with the standard line of the equator.

at right angles indicates no distortion. If the shape of the circle is distorted into an ellipse, but the area is the same as the circle and the two lines l and m intersect at angles greater or less than 90°, there is no areal distortion, but there is angular distortion.

A map showing multiple Tissot indicatrix circles is a valuable aid to determine projection distortion (see Figure 5.11). The revealed patterns of distortion help in choosing the appropriate projection for a particular area.

Combining GI from Different Projections

The large number of projections available means that great care must be taken when working with GI from different sources. Projections for GIS provide a great deal of flexibility, but easily introduce problems when working with data created using different projections. You should note that projections used for GI differ from maps in an important way. When a map is made, one single projection is used with a single scale for the entire map. The same thing applies for GI with one important difference: the coordinate system of the GI usually is much larger than a piece of paper used for a map. The GI must be scaled another time when a map is made, which can introduce some distortion. Obviously, if the GI is stored in the coordinates of a piece of paper, it is much harder to use it with other data, so this makes sense.

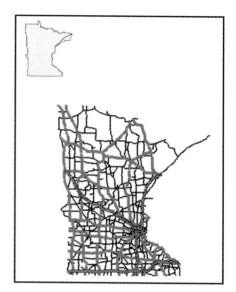

FIGURE 5.12. An example of an obvious error resulting from using data sources for the same area (Minnesota) but with different projections. Diagnosing the causes of such errors and resolving them can be very time-consuming if information about the choices made regarding the respective projections is unavailable.

Assuming that different data sets of GI for the same area use the same projection can lead to vast problems (see Figure 5.12). Usually the problems when combining GI from different projections are so obvious that they can't be missed. Sometimes the distortions are slight and may seem inexplicable: a road from one data source is 2 m away from the property that runs along it from another data source. If care is not taken, it is possible to create great errors by combining data prepared from different projections. The same concern applies to coordinate systems, the topic for the next chapter, where we will look at these issues in more detail.

Summary

Projections have been the core of cartography and the basis for representing GI. For millennia people have developed projections to find ways to represent the three-dimensional world humans live on in two dimensions—a format much better suited for recording observations and measurements. While it is possible to make maps without a projection, unprojected GI or maps are inaccurate and distorted for all larger areas and many small areas. A projection can be applied to different areas and at different scales. The smaller the area, the more accurate a projection can be. How accurate the projection is depends on how the projection is constructed and what underlying model of the earth's form it uses. Basic characteristics of a projection are its orientation, tangency, and form. Projections have several properties. The most important properties are the preservation of angles (conformality) and the preservation of areas (equivalent). Only one of these two properties can be preserved in any one projection. Some projections distort both properties and are called compromise projections. The resulting distortions can be ascertained and described using a Tissot indicatrix. Because of the number of differences, it is important to assess the characteristics and properties of projections when working with GI, especially when combining GI from different sources.

REVIEW QUESTIONS

1. Identify the type (equal angle, equal area, compromise) of the following projections:

    Mercator   Lambert   Mollweide   sinusoidal   azimuthal   Robinson

2. What is the difference between a secant and a tangent projection?

3. What is a transverse projection?

4. Why is a transverse Mercator projection better for north–south oriented areas and states (e.g., Illinois) than a Lambert conformal conic projection?

5. What are the three important characteristics of projections?

6. Why is most GI projected to a two-dimensional, Cartesian coordinate system?

7. Why should you never combine GI from different projections?

8. How can positional distortion be measured?

9. What is the difference between a geoid and a spheroid?

10. Why are Mercator and Peters projections technically satisfactory? Why do people consider the Mercator projection to be a bad projection?

ANSWERS

1. Identify the type (equal angle, equal area, compromise) of the following projections:

    Mercator   Lambert   Mollweide   sinusoidal   azimuthal   Robinson
    (Equal shape) (Equal area) (Equal area) (Equal area) (Equal distance) (Compromise)

2. What is the difference between a secant and a tangent projection?

    A secant projection surface "touches" the earth's surface in two places; a tangent projection "touches" only at one.

3. What is a transverse projection?

    A transverse projection is a cylindrical projection, which is normally oriented east–west, rotated 90 degrees to a north–south orientation.

4. Why is a transverse Mercator projection better for north–south oriented areas and states (e.g., Illinois) than a Lambert conformal conic projection?

The conic projection works best for areas with an east–west orientation; its line(s) of tangency run east–west. The transverse Mercator projection's line of tangency runs north–south, providing a more accurate positional reference than the Lambert conformal conic projection of the same area.

5. What are the three important characteristics of projections?

Equal shape: preservation of shapes; equal area: preservation of areas; equal distance; preservation of distances

6. Why is most GI projected to a two-dimensional, Cartesian coordinate system?

Several reasons need to be considered. Much GI comes from maps with such coordinate systems. Most GI is used to make planar maps. Most GIS are designed to store two-dimensional coordinate locations.

7. Why should you never combine GI from different projections?

GI from different projections for the same area will be in different coordinate systems that do not align properly.

8. How can positional distortion be measured?

For small-scale maps, Tissot's indicatrix provides a good graphical indicator. Large-scale maps, showing small areas, require the use of statistical measures.

9. What is the difference between a geoid and a spheroid?

A geoid is a more accurate representation of the earth's surface, accounting for local variations. A spheroid is a more round form that fails to account for local variations and the oblateness of the earth resulting from its spin.

10. Why are Mercator and Peters projections technically satisfactory? Why do people consider the Mercator projection to be a bad projection?

The Mercator projection is well suited for compass navigation at sea. The Peters projection is a compromise that offers a different way of representing the world. The overuse and ill-suited use of the Mercator projection to show regions of the world has led to the Mercator acquiring a bad reputation.

Chapter Readings

Jones, C. (1997). *Geographical information systems and computer cartography.* Upper Saddle River, NJ: Prentice Hall.

For a fascinating, if wide-reaching, biography and study of a person who was instrumental in determining the elliptical shape of the earth, see

Terrall, M. (2002). *The man who flattened the earth: Maupertuis and the sciences in the Enlightenment*. Chicago: University of Chicago Press.

For information about the basic mathematical principles of cartography, see

Cotter, C. H. (1966). *The astronomical and mathematical foundations of geography*. New York: Elsevier.

For a history of projections, see

Montgomery, S. (1996). Naming the heavens: A brief history of earthly projections. *Science as Culture, 5*(25), 546–587.

For a very thorough history of projections, see

Snyder, J. P. (1993). *Flattening the earth: Two thousand years of map projections*. Chicago: University of Chicago Press.

Web Resources

↻ More specific information about projection parameters and accuracy is provided by government agencies, for example, the California Department of Fish and Game: *www.dfg.ca.gov/biogeodata/gis/pdfs/DFG_Projection_and_Datum_Guidelines.pdf*.

↻ A good resource for fundamentals of geodesy is provided by the U.S. National Geospatial-Intelligence Agency, *Geodesy for the Layman*, available online at *www.ngs.noaa.gov/PUBS_LIB/Geodesy4Layman/toc.htm*.

↻ For information about homemade map projections using plastic bottles and the like, see *http://octopus.gma.org/surfing/imaging/mapproj.html*.

↻ This first of four articles offers a very well written and detailed discussion of the new U.S. datums: Minkel, D. H., & Dennis, M. L. (2012). Frames for the future: New datum definitions for modernization of the U.S. NSRS (Part 1 of 4). *The American Surveyor, 9*(1). Available online at *www.amerisurv.com/content/view/9609*.

EXERCISES

1. Projections for Different Needs

If you collect maps from magazines and newspapers for a few weeks, you will have a pretty sizable collection of different kinds of maps and different kinds of projections.

Come up with a list of different uses of maps and the projections used for each.

Think about how projections can preserve the shape of things on the earth, their size, or the distance from a point or along a line, or must compromise between these three projection properties.

Knowing what you do now about the different qualities of projections, what do you think about newspaper maps that do not indicate the projection? Are they common? What kind of errors do you think can arise?

If possible, you also can explore the collection of maps and atlases in a nearby library.

2. Questions for Map Projections

1. Is the map whole or broken up?
2. What shape does the projection make the map?
3. How are features (continents and islands) arranged?
4. Are gridlines curved or straight?
5. Do parallels and meridians cross at right angles?

EXTENDED EXERCISE

3. Sinusoidal Projection

Overview

In this exercise you will calculate values for a sinusoidal projection that you produce.

Concepts

The location of a point (x, y) in a sinusoidal equal area projection is calculated for this exercise in two steps. First, the longitude value is transformed to east–west values (x) by multiplying the longitude value times the radius and times the cosine of the latitude. Multiplying the longitude values by a cosine of latitude creates the gradually increasing distortion of areas farther away from the equator. The north–south values (y) of the projection are calculated through a linear relationship between the radius and the latitude. Second, you will scale the calculated x and y values to fit a map on a piece of paper by determining a scale ratio that transforms the radius of the sphere (6,371 km).

Exercise Steps and Questions

Preparation

In this exercise you will be calculating a projection of a graticule. You will have to do the calculations and show that you have done them, but you can work with other people to check your answers and determine the process. Before the calculating part of this exercise, let's look at the fundamental problems of projecting a spherical object on a plane.

Part 1. Angle Measures: Degrees and Radians

In Part 2 of this exercise, you will need to make the calculations in radians. Radians are one of three ways to measure angles. They are mainly used for engineering and science. We won't spend much time getting into the mathematics of angular measures. For this exercise, you only need to understand the relationship between degree and radian measures of angles.

If you know an angle measure in degrees, you can easily convert it to radians, another measure for angles used in engineering and scientific calculations:

$$radians = (degrees \cdot \pi)/180$$

For example, 180 degrees equals 3.14 radians; 90 degrees equals 1.57 radians; 45 degrees equals 0.785 radians. As the examples show, radians express angular measures in relation to the radius.

Part 2. Construct a Sinusoidal Projection of a Graticule

STEP 1: CALCULATE THE PROJECTION.

Use the table below for recording the results of your calculations. The rows indicating latitude are on the left and the columns indicating longitude are on the top. You will be calculating the sinusoidal projection for latitudes 0°, 30°, 60°, and 90°, and for longitudes 0°, 30°, 60°, 90°, 120°, 150°, and 180°. Your results will be in kilometers, or, for an idealized projection surface, about 10,000 km in length and height.

The equations you will use are:

$$x = radius \cdot longitude \cdot cosine (latitude)$$
$$y = radius \cdot latitude$$

where radius = 6,371 km. Remember: Convert all angle measures from degrees to radians by multiplying by pi and dividing by 180 degrees. For example, 30° corresponds to π/6 *using the conversion equation from above.*

Table of Projected Values (Step 1)

Latitude	Longitude						
	0°	30°	60°	90°	120°	150°	180°
0°	0,0						
30°							
60°							
90°							

STEP 2: SCALE THE *X, Y* VALUES AND THEN GRAPH THEM.

The *x, y* values calculated in Step 1 are in kilometers; therefore they are certainly too large to fit on a piece of paper. As with creating any other map, the values need to be converted to map units by determining a ratio that fits the x, y values on the sheet of paper you use (8.5 × 11 inches, or approximately 22 × 33 cm). Scale can be determined by putting the ground values and map values in the same units, here cm, and calculating the ratio between the shortest ground value distance and the longest map value distance.

Determine this value and fill it in here:

Scale factor: _____

With the scale factor, convert your original projected values to map units. Use the table below for those calculations.

Table of Projected Values (Step 2)

Latitude	Longitude							
	0°	30°	60°	90°	120°	150°	180°
0°	0,0							
30°								
60°								
90°								

Using a ruler, graph each coordinate pair on the x and y axis on a separate piece of paper. The graph should look like the northeastern quadrant of a sinusoidal projection. When this is completed, label the axis with tick marks that indicate the corresponding degree value from 0° to 90° latitude and 0° to 180° longitude. This is a map projected to a sinusoidal projection.

Questions

1. The sinusoidal projection is an example of an equal-area projection. What are the major differences between this type of projection and conformal projections?

2. Why do the x values lack two-dimensional scaling at 0° longitude in the sinusoidal projection?

3. What are the major differences between the Mercator and sinusoidal projections? How big is a pole in each projection?

4. Minneapolis/St. Paul is located at approximately −93° longitude, 45° latitude. What are the x, y coordinates in the sinusoidal projection?

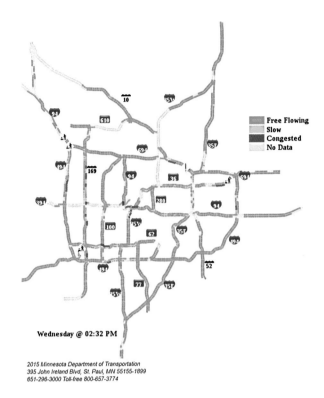

PLATE 1. The map showing current road status (events). The red–yellow–green sequence of colors that indicate levels of traffic density follows established graphic conventions in most of the world for indicating danger levels.

PLATE 2. Map showing the importance of conventions. Using blue for water and partially submerged ships follows commonplace concepts and associations.

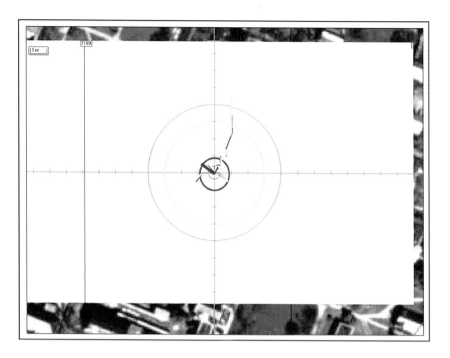

PLATE 7. Accuracy of GPS positioning with and without DGPS.

Two Garmin Etrex GPS units were placed at the intersection of two sidewalks at the center of the cross hairs, with a clear view of the sky.

The yellow circle represents a 15-meter radius of the expected accuracy of a unit without DGPS. The yellow line represents the position given by the unit during one hour of recording. Notice that most of the error occurred when the unit was first turned on. After a minute, accuracy was within 5 meters.

The green circle represents a 3-meter radius of the expected accuracy of a unit with a DGPS signal. The green line represents the track. Once again most of the error (but still within 3 meters) occurred when the unit was first turned on.

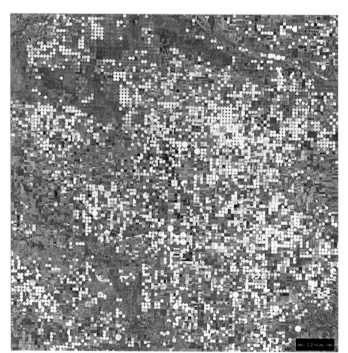

PLATE 8. Center-pivot irrigation systems create red circles of healthy vegetation in this image of croplands near Garden City, Kansas.

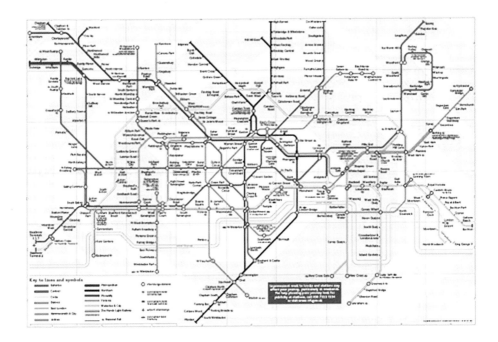

PLATE 9. The London Tube Map became the reference for easy-to-read maps of transportation networks in part because it only shows topological connections. Reprinted by permission of Transport for London.

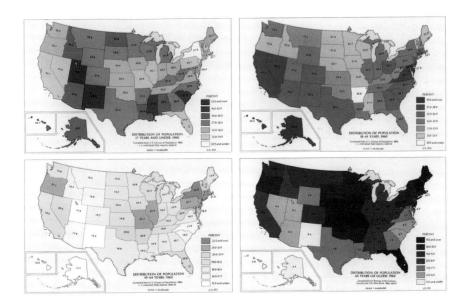

PLATE 10. Thematic maps are often used to create easily read comparisons.

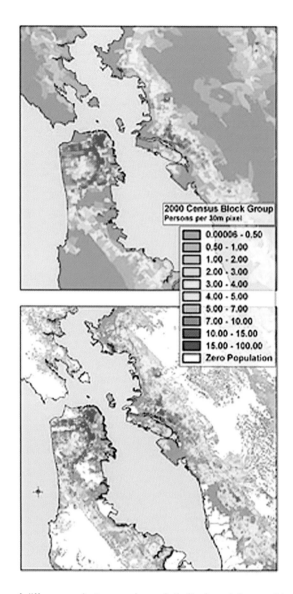

PLATE 11. Illustration of differences between choropleth (top) and dasymetric (bottom) mapping.

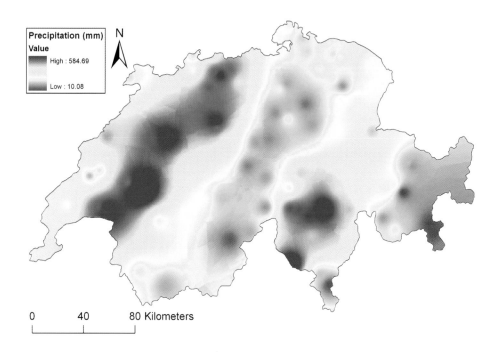

PLATE 12. Precipitation amounts based on a field model generated using inverse distance weighting.

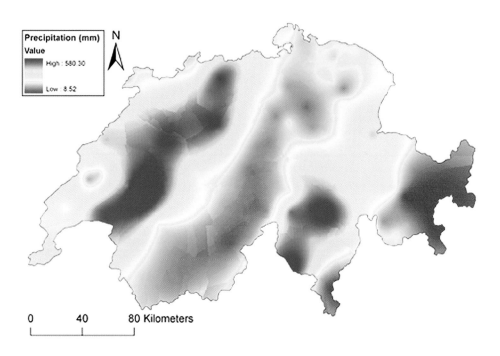

PLATE 13. Kriging (employing the same precipitation data used in Plate 12).

PLATE 14. A bicycle ride across the United States provides the subject matter for adding to Google Earth's rich selection of GI. A visualization tool for most people, it has become emblematic for the new directions that GIS is going in the eyes of many users.

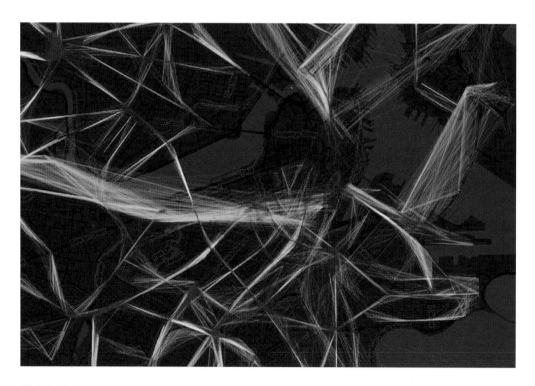

PLATE 15. A map from Andy Woodruff, Axis Maps, showing average public bus speeds in the Boston area.

CHAPTER 6

Locational and Coordinate Systems

Applying Projections

Projections make it possible to create maps and two-dimensional GI using locational and **coordinate systems** to create common reference systems. Often referred to collectively as "reference systems," they are used by national governments, state and provincial governments, local governments, the military, nonprofits, and businesses around the world. They make it possible to work with GI and maps without following the specifics of projections and datums. Their use is widespread, and thus they are a crucial reference. When different entities use a common reference system, this simplifies the recording of the location of things and events and makes it possible to combine information from different sources and verify distortions in positional measurements.

A commonly used locational or coordinate system helps greatly to minimize distortion. Accordingly, and just as with projections, working with GI or maps can require knowing which locational or coordinate system is used. Different locational or coordinate systems often give different locations for the same things and events. Even if they seem to overlap when drawn together, varying degrees of differences can lead to subtle or significant errors that impact analysis.

Locational systems are different from coordinate systems. Although the terms are often used interchangeably, it is important to recognize a key difference. A locational system *can* be referenced to a projection; a coordinate system *must* be referenced to a projection and *can* utilize a reference model that specifies the earth's shape and size. Usually this is known as a *datum*. Locational systems, generally with orthogonal coordinates, are only valid for particular data or possibly even with one map and may not have any connection to other locational and coordinate systems.

In this chapter you will read about the creation, history, and use of locational and coordinate systems. You will also find out about some issues in transforming GI between different coordinate systems. A more detailed discussion of related public administrative uses and issues is found in Chapter 13.

Locational Systems

Locational systems use a locally defined coordinate system or grid to indicate locations. Often they do not use projection or datum. These systems may be useful for some purposes in relatively large areas, but even then quickly run into accuracy problems due to the failure to consider the curved surface of the earth and lack of a clear reference to a geodetic control system. Their usefulness is also greatly limited by the degree of adoption. That is, a locational system for maps of city parks, state fairs, shopping centers, downtown, campgrounds, and so on work fine on one particular map, but if other people create another more popular locational system for the city, it may run into disuse and disregard.

To understand the significance of locational systems, we can begin with the Roman centuration across many areas of Europe, which is still geographically significant today. The practices of the Roman centuration, like those of Egyptian surveyors, offer fascinating insights into the historical roots and centrality of locational and coordinate systems. The technical details of the centuration system also highlight metrics that a number of societies today still rely on. The similarity between the Roman foot, at 29.57 cm, and the modern American foot, at 30.48 cm, or just about 0.9 cm different, is a great example that points to the persistence of common measurements (see Table 6.1).

Roman administrators actively surveyed conquered and politically associated areas undergoing integration into the empire. The survey created

TABLE 6.1. Common U.S. Surveying Measurements

1 link = 0.66 feet or 7.92 inches

1 pole or 1 rod = 25 links or 16.5 feet

1 chain = 100 links, 4 rods, or 66 feet

80 chains = 1 mile, 320 rods, 1,760 yards, or 5,280 feet

1 acre = 10 sq. chains, 160 sq. rods, 4,840 sq. yards, or 43,560 sq. feet

1 square mile = 1 section of land or 640 acres

Township = 36 sq. miles (36-mile-sq. sections)

These survey measurements are historical and archaic, but because of their legal nature these historical surveys are still valid. Current surveys generally use standard or metric measurements.

new rectangular subdivisions of land that could be more easily administered and awarded to army veterans as compensation for their years of service. Evidence of centuration can still be found in areas of Europe and North Africa (see Figure 6.1).

Centuration usually involved the creation of a local location system based on two orthogonal meridians. One meridian ran north–south, the other ran east–west. Based on this initial grid, the area was further subdivided into smaller and smaller units of land. Because the meridians were local and not tied to a projection and datum, each centuration was a locational system (see Figure 6.2).

The Public Land Survey

The Public Land Survey (PLS), also known as the Public Land Survey System (PLSS), is broadly similar in concept to the Roman centuration. It is used in most areas of the United States to survey land for recording ownership using a grid-like system (a similar system, the Dominion Land Survey, is used in large parts of Canada). It has become very influential on the landscape of the United States and has had many impacts related to governance and administration, which are examined in Chapter 13.

FIGURE 6.1. Archaeological finds show that the Roman survey in Great Britain corresponds to current landscape features in some areas. The Roman locational systems continue to impact current geography.

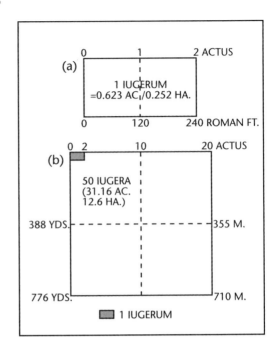

FIGURE 6.2. Hierarchical subdivision of the Roman centuration.

The PLS was created through the Land Ordinance of 1785 and the Northwest Ordinance of 1787, following the initiative of Thomas Jefferson and the support of other surveyors. The rationale was a plan to equitably provide access to land in the United States and help the government pay off its debts through the sale of land. After the Revolutionary War the U.S. government took on responsibility for all areas west of the original 13 states. The survey systems used prior to this time revealed themselves to have many problems that can still persist today. For example, the amount of land grants claimed in Georgia in 1796 was more than three times greater than the actual amount of land in the state.

All these western lands were considered to be the "public domain," except for the beds of navigable bodies of water, national installations such as military bases and national parks, and areas such as land grants that had already passed to private ownership prior to subdivision by the government. The latter included land awarded to private individuals by the governments of France, Mexico, and Spain. Part of the original intention was to efficiently allocate land to soldiers who had fought for the United States, but the PLS was also seen to be a way to help pay off debts from the war and to cover future expenses. The original public domain included the land ceded to the federal government by the thirteen original states, supplemented with acquisitions from Native Americans and foreign powers. It encompasses major portions of the land area of 30 contemporary southern and western states. Almost 1.5 million acres were surveyed into the PLS system of townships, ranges, and sections (see Figure 6.3).

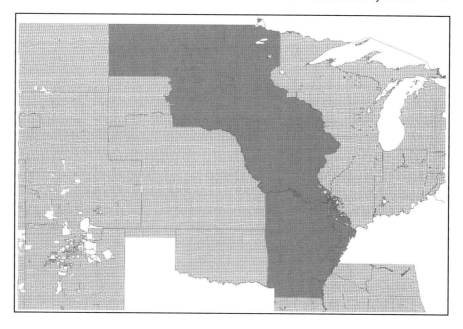

■ FIGURE 6.3. Area of the U.S. PLS surveyed from the Fifth Meridian.

The PLS is a hierarchical land subdivision system that makes it possible to locate land. The hierarchy begins with two of the 34 meridians, which are distinct and unrelated to the other PLS meridians. The 34 principal meridians run north–south. Each is named, and allows the identification of different surveys. The meridians meet baselines, as they are called, which run east–west and are perpendicular to a principal meridian. The first subdivision of theoretically 6 × 6 mile units is organized by *townships*, which indicate the location north or south of a baseline, and *ranges*, which indicate the location east or west of a principal meridian. Each 6 × 6 mile unit is called a township and is further theoretically divided into 36 equal 1 × 1 mile sections. Each section has a theoretical area of 640 acres and can be further divided into aliquot parts including half sections, quarter sections, and quarter-quarter sections (see Figure 6.4). The location of all land in this system consists of the state name, the name of the principal meridian, township and range designations with cardinal direction, and the section number.

In some areas, due to survey difficulties, errors, or even fraudulent surveys, the townships and sections may vary considerably from the theoretical system of 6 × 6 mile townships and 36 equal 1 × 1 mile sections. (See Figure 6.4.) Fraudulent surveys, surveys that were indicated as correct, but in actuality contained some discrepancy, were often detected before becoming legally binding, but in some cases frauds slipped through and became legally valid. In some other cases, surveys may have been conducted in difficult terrain, beyond the capabilities of the surveying equipment or the surveyors (see Figure 6.5). Errors may have been made in spite of thorough

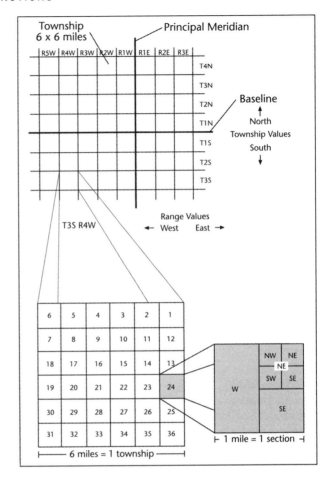

FIGURE 6.4. Nested hierarchical organization of the U.S. PLS.

cross-checking of survey measurements. Some people hired to conduct surveys sought short-cuts to simplify the work, and produced surveys that were complete on paper, but may have distorted things on the ground. Regardless of the limitations, errors, or frauds, the markers placed by the "original" surveys are considered the ultimate authority for all later land subdivisions and locations, even if some quite complex problems require later resolution.

The PLS's consequences go beyond the creation of a system for subdividing land and the development of the ability to systematically locate land in most of the United States (see Figure 6.6). The "original" survey marked the land for future development. If you have ever flown across the midwestern or western United States and looked out the window, you probably noticed a landscape that looks like a grid stretching out to the horizon. Traveling in a car on many roads in this area, you probably noticed that the road goes straight for a long time, with only minor deviations, and intersections with roads are mostly at right angles. Both are the consequences of the PLS. But there are more significant consequences, which are environmentally and economically significant. First, the PLS subdivision of land does not follow

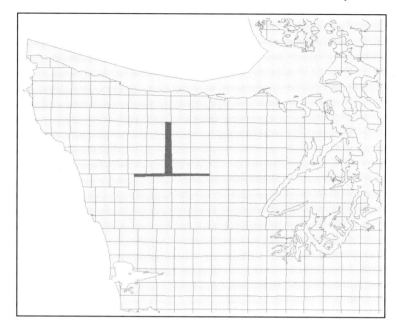

FIGURE 6.5. PLS in western Washington State highlighting unusual 1/2 townships and ranges. These peculiar townships are the results of difficulties surveying in the alpine Olympic Mountain Range. Other distortions are also plainly visible.

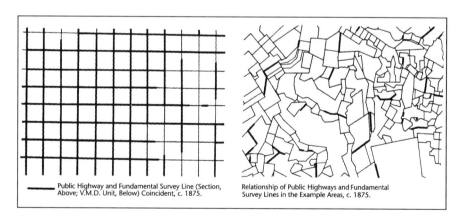

Public Highway and Fundamental Survey Line (Section, Above; V.M.D. Unit, Below) Coincident, c. 1875.

Relationship of Public Highways and Fundamental Survey Lines in the Example Areas, c. 1875.

FIGURE 6.6. Different lengths of roads mean different lengths of bridges between systematic and nonsystematic surveyed areas.

existing natural features, which usually help guide the use of land. Intensive farming practices in PLS areas can more easily have detrimental effects than the same practices in areas surveyed using metes-and-bounds surveys. The roads in the PLS may be easier to drive on, but the costs of maintaining bridges may be higher because they have to be longer. In Norman Thrower's study, he established that in an area surveyed using PLS 60% of all bridges were longer than 20 ft (6 m), versus only 20% in areas that were surveyed unsystematically (see Figure 6.7).

Local Coordinate Systems

A system like the PLS is registered to a meridian and a baseline that have known coordinate values. Further, any PLS locations can be associated with other coordinate systems. While it is possible to determine coordinate locations in the PLS, the system functions without any reference to coordinates associated with the earth's size or shape.

Local systems are further removed from relationships with the earth's size or surface (see Figure 6.8). Although the Roman centuration relied on meridians, which were surveyed based on astronomical observations and measurements, the survey of PLS meridians makes the relationship with the earth's size and surface and with other meridians secondary, meaning for surveying and legal purposes that a portion of the PLS is essentially a local location system. Smaller local coordinate systems are commonplace because they are handy for quickly aiding people using and orienting themselves with maps. However, they are of no use for recording the location of things and events when they should be used with other locational and coordinate systems.

Aerial photograph of an area in northwestern Ohio subdivided in the manner of the United States Land Survey System. The black x's indicate a quarter section (1/2 x 1/2 mile).

Aerial photograph of an area in the Virginia military district of Ohio subdivided in an unsystematic manner.

FIGURE 6.7. Different location systems used in surveying have significant geographical and environmental consequences.

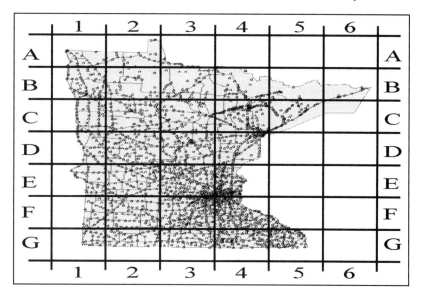

FIGURE 6.8. Figure showing an arbitrary local locational system for Minnesota highways and towns.

Rectangular Coordinate Systems

Rectangular coordinate systems are different from locational systems in that they are associated with a particular model of the earth's size and shape (geoid or ellipsoid). A datum is usually also associated with a particular projection. However, this broader use of the term has become more commonplace.

The PLS system, when associated with a rectangular coordinate system, takes on characteristics of both locational and coordinate systems, although surveyed locations never replace the legally binding locational system. In other words, for mortgage lenders, title insurances, and banks, the locational system description is what is important—the associated coordinates have little significance.

Metes-and-Bounds Systems

Metes-and-bounds systems, used historically most often for local maps that record the location of parcels, were often connected to local meridians and parallels. They can be associated with a projection and made into rectangular coordinate systems with great ease. Most metes-and-bounds systems are nowadays connected to a rectangular coordinate system. In the United States, metes-and-bounds is the legally recognized system for recording parcels in the areas of the original 13 colonies and Texas, plus the areas of a few other states. In most areas of the world, metes-and-bounds systems are the more common systems for recording not only the extents of land parcels, but also of legally registering land ownership, rights, and responsibilities. The metes-and-bounds system can either start with recognized origin points and then survey the boundaries of parcel boundaries based on distance and angle relationships, or just survey boundaries based on existing surveys. The former is the preferred approach, as it avoids many inaccuracies that lead to significant land conflicts.

Metes refer to the distances and angles. *Bounds* refer to the corners and points that define the outline of the surveyed area. A metes-and-bounds description is a narrative that describes the clockwise or counterclockwise path around the perimeter. A simple example of a metes-and-bounds description can read like this:

> Beginning from the southwest corner of section, thence north 1,320 feet; thence east 1,735 feet to the true point of beginning thence east 500 feet, more or less to State Road 35 right-of-way, thence northwesterly along said right-of-way.

A more detailed metes-and-bounds description can also describe the vicinity of the surveyed area, exempted areas, and additional rights to areas described in the survey. The following is an example of a more detailed description of the diagram in Figure 6.9. Once facing due south or due north, the numbers in the parentheses are the offset angle (in degrees, minutes, and seconds west or east) using bearings of the surveying equipment.

> Beginning at the concrete monument, thence S (83 deg 58′06″W) for 211.19 ft along the North right-of-way of the highway; thence N (18 deg 40′10″E) for 150.00 ft along the East line of Brown; thence S (72 deg 21′10″E) for 170.00 ft

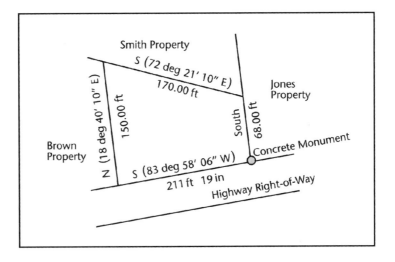

FIGURE 6.9. Simplified example of a metes-and-bounds description.

along the South line of Smith; thence South for 68.00 ft along the West line of Jones, back to the point of beginning.

State–Plane Coordinate System

The State–Plane Coordinate System (SPCS) is a system for specifying positions of geodetic stations and measuring location using plane rectangular coordinates in the United States. This is a local coordinate system for each state which is legally defined in federal and state law. This coordinate system divides all fifty states of the United States, Puerto Rico, and the U.S. Virgin Islands into 124 numbered sections, referred to as "zones." Larger states generally have multiple SPCS zones—for example, Minnesota has three and California has six. Each zone has an assigned code number that defines the projection parameters for the region. SPCS uses three projections, depending on the orientation of the zone and the state. The Lambert conformal conic projection is used for areas with an east–west orientation. Areas with a north–south orientation use a transverse Mercator projection. The Alaskan panhandle uses an oblique Mercator projection.

The SPCS uses two datums, the North American Datum of 1927 (see Figure 6.10) and the North American Datum of 1983 (NAD 1927 and NAD 1983, respectively) based on different models of the earth's shape and size. NAD 1927 uses Clarke's 1866 spheroid (equatorial radius 6,378,206, flattening 1/294.98); NAD 1983 uses GRS 1980 (equatorial radius 6,378,137, flattening 1/298.26). The differences between the datums' geoids are significant and lead to sizable differences (up to several hundred meters) between locations recorded using SPCS NAD 1927 and SPCS NAD 1983 (see Chapter 3). Since then, numerous regional modifications have also been made. These changes necessitate great care when working with GI from the United States. GI collected with the later NADCON or HARN improvements will differ from NAD 1983 data (see Table 6.2).

FIGURE 6.10. State–Plane Coordinate System (SPCS) zones using the North American Datum 1927 (NAD 1927).

TABLE 6.2. The Parameters for the Three SPCS Zones in Minnesota Using the 1927 North American Datum[a]

SPCS Zone	Semimajor Axis (m)	Semiminor Axis (m)	Southern Standard Parallel	Northern Standard Parallel	Longitude of Origin	Latitude of Grid Origin
North	6378206.4	6356583.8	47 02 00	48 38 00	−93 06 00	46 30 00
Central	6378206.4	6356583.8	45 37 00	47 03 00	−94 15 00	45 00 00
South	6378206.4	6356583.8	43 47 00	45 13 00	−94 00 00	43 00 00

[a]See *http://rocky.dot.state.mn.us/geod/projections.htm#27MSP* for further information.

The U.S. National Grid

The U.S. National Grid (USNG) was standardized in 2001 in response to needs for a single coordinate system for the entire United States, especially for location-based services for mobile phones, GPS, and other navigation devices. The USNG can be extended to include coordinates for locations anywhere in the world. It is not intended to replace the SPCS or other coordinate systems, but it provides a coordinate system with national scope (see Figure 6.11). It is a hierarchical system, using the grid system specified in the Military Grid Reference System (MGRS). The coordinates are also identical with UTM coordinates in areas of the United States. Coordinates can be specified in two precisions. For example, the location of the Washington Monument in Washington, DC, is:

General reference:	18SUJ23480647—precision 10 m
Special application:	18SUJ2348316806479498—precision 1 mm

The U.S. geographic area is divided into 6-degree longitudinal zones designated by a number and 8-degree latitudinal bands designated by a letter. Each area receives a unique alphanumeric Grid Zone Designator (GZD)—for example, 18S. Each GZD 6 × 8 degree area is divided into a systematic scheme of 100,000-m squares where a two-letter pair identifies each square—for example, UJ. A point position within the 100,000-m square shall be given by the UTM grid coordinates in terms of its easting (E) and northing (N). The number of digits specified the precision:

18SUJ20	Locates a point with a precision of 10 km
18SUJ2306	Locates a point with a precision of 1 km
18SUJ234064	Locates a point with a precision of 100 m
18SUJ23480647	Locates a point with a precision of 10 m
18SUJ2348306479	Locates a point with a precision of 1 m

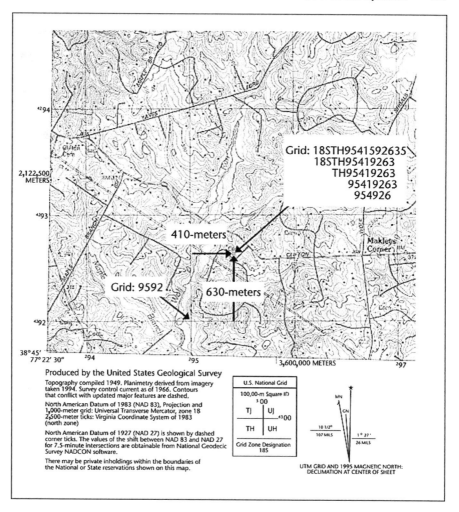

Produced by the United States Geological Survey

Topography compiled 1949. Planimetry derived from imagery taken 1994. Survey control current as of 1966. Contours that conflict with major features are dashed.

North American Datum of 1983 (NAD 83), Projection and 1,000-meter grid: Universal Transverse Mercator, zone 18 2,500-meter ticks: Virginia Coordinate System of 1983 (north zone)

North American Datum of 1927 (NAD 27) is shown by dashed corner ticks. The values of the shift between NAD 83 and NAD 27 for 7.5-minute intersections are obtainable from National Geodecic Survey NADCON software.

There may be private inholdings within the boundaries of the National or State reservations shown on this map.

FIGURE 6.11. Example of a location in the USNG coordinate system.

Universal Transverse Mercator

The Universal Transverse Mercator (UTM) grid was developed in the 1940s by the U.S. Army Corps of Engineers. In this coordinate system, the world is divided into 60 north–south zones, each covering a strip 6° wide in longitude (see Figure 6.12). These zones are numbered consecutively beginning with Zone 1, between 180° and 174° west longitude, and progressing eastward to Zone 60, between 174° and 180° east longitude. The conterminous 48 United States are covered by 10 zones, from Zone 10 on the West Coast through Zone 19 in New England. In each zone, coordinates are measured north and east in meters. The northing values are measured continuously from zero at the equator, in a northerly direction. To avoid negative numbers for locations south of the equator, the equator has an arbitrary false northing value of

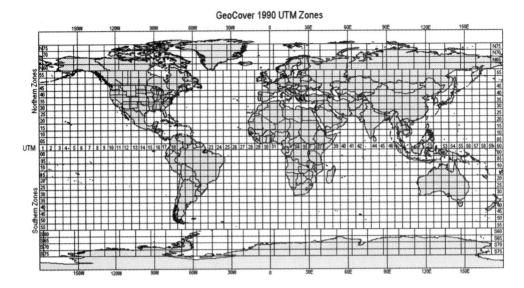

FIGURE 6.12. UTM zones is a widely used coordinate system based on usage of the Transverse Mercator projection in a global system of 6° strips from pole to pole.

10,000,000 meters. A central meridian through the middle of each 6° zone is assigned an easting value of 500,000 meters. Grid values to the west of this central meridian are less than 500,000; to the east, more than 500,000.

Other National Grids

Most countries in the world have one or more national grids, analogous in function to the U.S. system. In the world there are thousands of these systems. The following examples are exemplary for different approaches to organizing coordinates. The United Kingdom, for example, has a hierarchical system that begins with a grid of 100 100-km cells, identified by two letters. Each 100-km cell is further divided into 100 10-km grid cells. A 10-km grid cell is further divided into 100 1-km grid cells. Germany uses a system similar to the UTM, but bases it on 3°-wide stripes at the 6, 9, 12, and 15 meridians. The zones are numbered two through five, or the meridian longitude divided by three. A false easting of 500,000 meters is calculated for east–west coordinates in each stripe, and north–south coordinates are the distance to the equator. North–south coordinate values have seven digits and east–west coordinate values have six digits, precluding switching the coordinates. Australia has developed new national grids directly as coordinate systems at frequent intervals, reflecting both frequent tectonic movement (up to 7 cm/year) and improvements in geoid measurements. The Geocentric Datum of Australia coordinate system is the most recent and is based on the ellipsoid measurements from GRS 1980 and coordinates from the International Earth Rotation Service.

Smaller countries generally use only one projection and geoid for the entire country. People who live here (or use only local maps) may never even have to learn about projections and be concerned with how to combine data from different projections.

Polar Coordinate Systems

Polar coordinate systems are necessary in areas around the poles, but can be used for specialized applications in other areas as well. A two-dimensional polar coordinate system records locations based on an angle measurement (azimuth) from the central point, the pole, of the coordinate system and a distance to that point (see Figure 6.13a). A three-dimensional coordinate system records location with two angle measurements and the distance to the measured point from the center. One angle measurement records the horizontal angle on the XY plane, the other records the angle on the Z plane (see Figure 6.13b).

Spherical Coordinate Systems

A basic spherical coordinate system records the location of things and events using three values: x, y, and z. The value x stands for the east–west coordinate value, y for the north–south coordinate value, and z for the elevation in relation to a reference height. It is similar to a three-dimensional polar coordinate system, except that the origin point lies at the center of the coordinate space—for example, the theoretical center of a sphere.

A two-dimensional plane of x, y coordinates that correspond to latitude and longitude coordinates is often used by GIS to represent the entire world at once, but it introduces such grave distortions that it should be avoided.

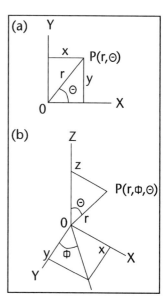

FIGURE 6.13. Two-dimensional polar coordinate systems.

Global grids follow a different approach to creating a global grid, usually based on hexagons or octahedrons, to subdivide a sphere hierarchically into smaller and smaller triangular facets (see Figure 6.14). These coordinate systems are still rather uncommon—they are mainly used for satellite tracking and studying global processes—although the advantages of these systems are significant. For measurements of distance and area at continental and global scales, the results of making measurements using almost any projection are quite unsatisfactory.

Scales and Transformations

Any geographical map you will ever see has a scale. It may be only implicit, as in a graphic artist's rendering of a summer festival site, or a city's advertising map, but more often you'll find explicit scales. An important question for the use and creation of GI and maps is: What is the appropriate scale? A scale too small, that shows a large area, will require that small specific things and events be removed, whereas a large scale may lead to important contextual information being left out. To work well with scale it is critical to familiarize yourself with different ways of representing the relationship between a distance unit of GI on a map and the corresponding distance unit on the ground.

Scale is shown for GI and maps in three ways (see Figure 6.15):

- Representative fraction
- Scale bar
- Statement

The three types are equivalents, but have different representations. A *representative fraction* provides a ratio between the same units of measure on a page and on the ground. A *scale bar* graphically represents distinct

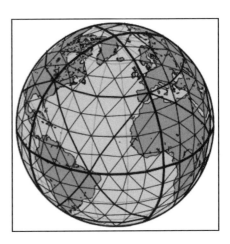

FIGURE 6.14. A global tessellation based on hexagons.

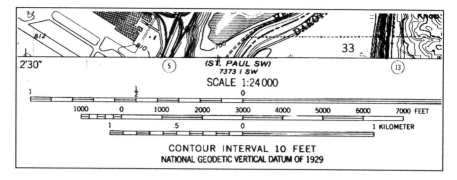

Representative scale and scale bars from a USGS map.

distances at the scale of the GI or map. A *statement* describes the scale in words. The most important thing for representing scale is that the measurement units on the page (or for the GI) and on the ground must be kept the same. For example, the representative fraction scale 1:24,000 indicates that 1 inch on the map corresponds to 24,000 inches on the ground. Divide by 12 (the number of inches in a foot) and you'll have the basis for this alternative statement of the map's scale: "1 inch equals 2,000 feet." Using metric units, the calculations are even easier: the representative fraction scale 1:25,000 indicates that 1 cm on the map corresponds to 25,000 cm on the ground. Divide by 100,000 (the number of cm in a km) to determine the statement of scale "1 cm equals 250 m or a quarter kilometer."

Scale Transformations

GI, whether collected in the field, collected from existing GI, or digitized from existing maps, can be readily transformed to other scales. The scaling of GI may be helpful for many reasons. Most often, scale transformations allow the association of any arbitrary coordinates from known places—for example, building corners or street intersections—to be associated with coordinates of the same places in other coordinate systems. In this way, locations of things and events drawn on a piece of paper can be transformed into GI using a coordinate system (see Table 6.3).

Scale transformations allow for an infinite number of alterations to shapes and changes. They can change all axes by the same factor, each axis by different factors, locally vary the transformation values, or use logarithmic factors. These different types of scale transformations are necessary to support the different types of changes to coordinates required when working with GI from different sources.

Several things need to be considered for working with scale transformations. First, it is important to remember to keep using the same units throughout the transformation. GI locations stored in metric units should be kept in metric units. If a transformation is made between metric and standard units, be sure that all GI was converted using the same constants. The

TABLE 6.3. Representative Scale and Equivalent Ground Distances

Scale	Ground distance
Standard (inches)	
1:2,400	200 ft
1:20,000	1,667 ft
1:24,000	2,000 ft
1:62,500	approximately 1 mile
1:63,360	5,280 feet (exactly 1 mile)
1:125,000	approximately 2 miles
1:800,000	approximately 12.6 miles
Metric (centimeters)	
1:1,000	10 m
1:2,500	25 m
1:10,000	100 m
1:25,000	250 m
1:50,000	500 m
1:100,000	1,000 m (1 km)
1:250,000	2,500 m (2.5 km)
1:500,000	5,000 m (5 km)
1:1,000,000	10,000 m (10 km)
1:2,000,000	20,000 m (20 km)

transformations can also alter geographic representations and cartographic representations, leading to GI that is not only inaccurate but also incorrect. A common example is scaling small-scale maps to match large-scale maps of the same area. Because the small-scale maps lack accuracy in comparison to a large-scale map, differences between the two maps can be the results of changes made during the generalization process—for example, when a road is displaced to fit a railroad track symbol next to a bend in a river.

A Sample Scale Transformation

The simplest type of scale transformation is an affine transformation (see Figure 6.16). Even an affine transformation makes it possible to scale, rotate, skew, and translate GI coordinates, both in projection transformations and in georeferencing.

Affine transformations use two equations for the x and y coordinates of two-dimensional GI.

$$x' = Ax + By + C$$
$$y' = Dx + Ey + F$$

Accuracy of Georeferencing

According to an article in the *International Journal of Geographic Information Science*, six steps are required for determining georeferences to places lacking precise coordinate references—for example, 6 miles NW of Timmons, NV. Knowing these six steps is important because many places with descriptive data lack accurate locational references; however, coordinates must be used for storing the descriptions as GI. The point–radius method summarized here provides consistent and accurate interpretations of locality descriptions and identifies potential sources of uncertainty.

Step 1: Classify the locality description. The quality of the description should be assessed and classified. Only somewhat accurately described localities should be georeferenced.

Step 2: Determine coordinates. Coordinates can be retrieved from gazetteers, geographic name databases, maps, or from local descriptions with coordinates—for example, field notes with GPS coordinates. The numerical precision of coordinates should be preserved during processing to minimize the propagation of error. Next, *identify named places and determine their extents.* Every named place has an extent. This should be determined in the same manner as the coordinates of the locality. Most named places have a geographic center (courthouse, church) which should be used as the origin of circle defining the extent. Then, *determine offsets.* Many localities are located by their relation to another place—for example, 6 miles NW of Timmons. The direction from the place can usually be inferred, considering environmental constraints and additional information in the description. Supplementary sources are helpful.

Step 3: Calculate uncertainties. In this article, Wieczorek et al. consider six sources of uncertainty:

1. *Extent of the locality.* The maximum extent of two places in the locality is the maximum uncertainty.

2. *Unknown datum.* The differences can be as large as 500 m between NAD 1927 and NAD 1983. Theoretically the difference could be as large as 3,552 m.

3. *Imprecision in distance measurements.* Treat the decimal portion of distance measurements as a fraction and multiply the distance measurement by this fraction. Multiples of powers of 10 should be multiplied by 0.5 to that power of 10.

4. *Imprecision in direction measurements.* Translate cardinal directions to their degree equivalents, using half of that degree equivalent as the uncertainty.

5. *Imprecision in coordinate measurements.* Consider latitude and longitude error.

$$uncertainty = lat\_error2 + long\_error2$$

6. *Map scale.* Take the error of a map to be 1mm. For example, the uncertainty for a map of scale 1:500,000 is 500 m.

Step 4: Calculate combined uncertainties. The uncertainties without directional imprecision and combined distance and direction uncertainties should be calculated following map accuracy guidelines for the maps used. Distance uncertainties should take directional imprecision into account.

Step 5: Calculate overall error. Assuming a linear relationship between individual errors and total error, use a root-mean-square equation and apply the law of error propagation to determine the maximum potential error.

Step 6: Document the georeferencing process. Documentation of the process and considerations used in determining the georeferencing are important for people working with the locality information later.

Based on J. Wieczorek, Q. Guo, et al. (2004). The point–radius method for georeferencing locality descriptions and calculating associated uncertainty. *International Journal of Geographical Information Science, 18*(8), 745–767.

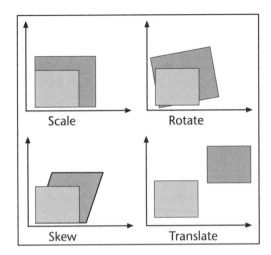

FIGURE 6.16. Affine scale transformation operations (generalized).

The values x and y stand for the coordinates of the input GI; x' and y' stand for the coordinate values of the transformed GI. A, B, C, D, E, and F are the six geometric parameters for transforming the GI coordinate values. Some GIS require the entry of these parameters; others will calculate them for you based on common reference points in the input GI and in the output GI. A linear transformation simply multiplies the coordinate values by the scale factor to obtain the scaled GI (see Figure 6.17).

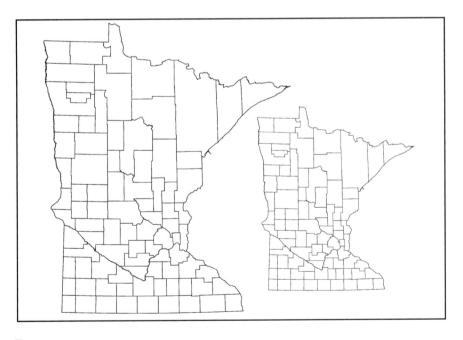

FIGURE 6.17. Map showing counties of Minnesota before (left) and after (right) scale transformation.

Summary

This chapter focuses on location systems and coordinate systems, both of which often involve a projection, but may be developed without any reference to the earth's size or shape. Location systems are more likely to be locally developed ways for describing location using a grid of letters and numbers. Coordinates without a reference to the earth's size and shape are a type of location system. Coordinate systems may have a reference to the earth's size and shape through a projection, generally described as a datum. Location systems are important because they are very common and can be used to coordinate activities. Land subdivision is an important activity involving both location systems and coordinate systems. It establishes the divisions of land used in determining ownership. The U.S. PLS is possibly the most widely used systematic survey for subdividing land. More common in the rest of the world are unsystematic surveys that use metes-and-bounds approaches to recording the boundaries of land parcels. Because of their importance, law often specifies coordinate systems; usually these are called national grids. In the United States, the SPCS is the best example. The newer USNG is an example of a more recent coordinate system that aims, among other things, to improve coordination among first responders by assuring a common reference system is used for locations. The older UTM system is widely used around the world, and thus retains great importance for working with GI, including some online mapping applications and virtual globes. All location systems and coordinate systems use a scale to reduce the size of measurements on the ground to map size or a comparable size in GI. GI from one scale can be easily converted to another scale; it is much more labor-intensive to scale different maps.

REVIEW QUESTIONS

1. What are common applications for spherical coordinate systems?

2. What is the main practical importance of coordinate systems?

3. What is the main difference between coordinate and locational systems?

4. What is the transformation from x, y to x', y' called when all scale factors are the same?

5. What is the difference between rectangular and polar coordinates?

6. For what purpose was Roman centuration devised?

7. What is the similarity between metes-and-bounds and the PLS in the United States?

8. What is the State–Plane Coordinate System?

9. How are locational and coordinate systems used for public administration?

10. Why are 3-D coordinate systems still uncommon?

11. Why do projection parameters need to change?

12. For the region or county you live in, what would be the ideal projection?

ANSWERS

1. What are common applications for spherical coordinate systems?

   Spherical coordinate systems are commonly used for satellite tracking and global models.

2. What is the main practical importance of coordinate systems?

   Coordinate systems provide for the common recording of positional locations against which distortions can be measured.

3. What is the main difference between coordinate and locational systems?

   Coordinate systems are mathematically defined based on a model of the earth's surface and shape; location systems can be mathematically defined, but are usually created without relating them to a model of the earth's surface and shape.

4. What is the transformation from x, y to x', y' called when all scale factors are the same?

   A constant scale transformation is called a linear transformation.

5. What is the difference between rectangular and polar coordinates?

   Rectangular coordinates are orthogonal, that is, the x, y origin is defined by a right angle.

6. For what purpose was Roman centuration devised?

   Colonizing and developing conquered areas.

7. What is the similarity between metes-and-bounds and the PLS in the United States?

   In the areas they respectively dominate, they are legally accepted means of recording land ownership.

8. What is the State–Plane Coordinate system?

   The SPCS was established during the 1930s in the United States to specify coordinate systems for each area. Many states have legally adopted the SPCS.

9. How are locational and coordinate systems used for public administration?

   Locational and coordinate systems are used for recording locations, helping government/private coordination, and providing a structure for future activities.

10. Why are 3-D coordinate systems still uncommon?

    The complex mathematics and lack of a common reference standard along with the abundance of 2-D maps and GI hinder the widespread use of 3-D coordinate systems.

11. Why do projection parameters need to change?

The earth's shape is constantly changing and the measurements of location have improved greatly. These are the main reasons for changing projection parameters, but regional and national changes can also influence the choice of new projection parameters.

12. For the city or county you live in, what would be the ideal projection?

Consider the extent of the area, its size, and the detail required. The "ideal projection" can be a question of its fitness for you.

Chapter Readings

Caravello, G. U., & Michieletto, P. (1999). Cultural landscape: Trace yesterday, presence today, perspective tomorrow for "Roman Centuriation" in rural Venetian territory. *Human Ecology Review, 6*(2), 45–50.

Dilke, O. A. W. (1985). *Greek and Roman maps*. London: Eastern Press.

Ferrar, M. J., & Richardson, A. (2003). *The Roman survey of Britain*. Oxford, UK: Hedges.

Goodchild, M. F., & Proctor, J. (1997). Scale in a digital geographic world. *Geographical and Environmental Modelling, 1*(1), 5–23.

Linklater, A. (2002). *Measuring America: How an untamed wilderness shaped the United States and fulfilled the promise of democracy*. New York: Walker & Company.

Thrower, N. J. W. (1966). *Original survey and land subdivision*. Chicago: Rand McNally.

Web Resources

↻ NOAA's National Geodetic Survey maintains key resources for locational and coordinate systems at *www.ngs.noaa.gov*.

↻ The documentation of the U.S. SPCS prepared by James Stem is available from NOAA at *www.ngs.noaa.gov/PUBS_LIB/ManualNOSNGS5.pdf*.

↻ A list of NGS publications on horizontal reference systems can be found at: *www.ngs.noaa.gov/PUBS_LIB/pub_horiz.shtml*.

↻ Roman centuration and surveying is described nicely by Richard Hucker in *www.fig.net/pub/fig2009/papers/hs01/hs01_hucker_3471.pdf*.

↻ The National Atlas provides an introduction to the PLSS at *http://nationalatlas. gov/articles/boundaries/a_plss.html*.

↻ Detailed instructions for the PLS are contained at *www.blm.gov/cadastral/ Manual/73man/id1.htm*.

↻ A very thorough and well-developed introduction to the history and legalities of land surveys and information in the United States is available at *www. premierdata.com/literature/Intro%20Land%20Information.pdf*.

↺ A good description of the metes-and-bounds survey system from Tennessee is available at *www.tngenweb.org/tnland/metes-b.htm*.

↺ Detailed description and parameters for the SPCS are available at *www.ngs. noaa.gov/PUBS_LIB/ManualNOSNGS5.pdf*.

↺ Resources on U.S. datum measurements and conversions are available at *www.ngs.noaa.gov/PC_PROD/pc_prod.shtml*.

↺ The U.S. Forest Service and Bureau of Land Management maintain a website for information for specific PLS questions at *www.geocommunicator.gov*.

↺ The Information and Service System for the European Coordinate Reference Systems (CRS) has a website at *www.crs-geo.eu*.

↺ Math on a Sphere | An Interactive Exploration of 3D Surfaces for Public Audiences is available at *www.crs-geo.eu* and *http://mathsphere.org*.

↺ How maps can deceive us by distorting space is available at: *http://geography. about.com/od/understandmaps/a/How-Maps-Can-Deceive-Us.htm*.

EXERCISE

1. Use PLS Coordinates

On a topographic map of the place you're from or one you're familiar with in the United States, determine the location of your home, school, and other important local features using the PLS coordinates. Townships should be given on the east and west edges of the map, ranges on the north and south.

EXTENDED EXERCISE

2. Locational and Coordinate Systems

Overview

In this exercise you will interpret the impact of land subdivision systems and learn how to read coordinates from United States Geological Survey (USGS) topographic maps. This exercise also introduces you to basic topographic mapping concepts and how to recognize them.

Concepts

Topographic maps are the most general purpose maps in circulation. They represent a multitude of features and relationships that you can "read" by looking at and studying a map. Among other things, they are useful for studying general land use development.

Exercise Steps and Questions

Step 1. Find a USGS 7.5 Minute Topographic Quadrangle

In this step you should use an index of topographic quads for a state in the United States to find a map of your home or a place you are familiar with. The feature you look for can also be a particular place, mountaintop, radio antenna, or structure—for example, a lighthouse. It should be small enough to be located distinctly: a small building or pond is OK, but not a large lake or structure. Most of all, *it should be someplace you are familiar with*. If you can't find the 7.5 quad you need, first check to see if an older quad is available for the area. If not, choose a feature somewhere else.

Answer these questions before continuing:

1. What is the name of the place you chose?
2. What is the type of feature?
3. Why did you choose it?
4. What is the latitude and longitude of the southeast corner of the map?
5. What is the distance, in kilometers and miles, from east to west across the map?

    Kilometers _____

    Miles _____

6. What is range of elevation in this area?

    Highest _____

    Lowest _____

    Average _____

7. Does this map show elevation in feet or meters?
8. Do you see *any* consequences of land subdivision—for example, in the orientation of roads?
9. What is the name of the map you chose?

Step 2. Locate Your Feature

10. Find the feature you described in question 2 and locate it using the following coordinate and land subdivision systems:

    Latitude _____   Longitude _____

    UTM Northing _____   UTM Easting _____

    Township _____   Range _____   Section _____

    You should use a straightedge for these measurements and interpolate the distance.

11. What is the datum of the UTM coordinates?

CHAPTER 7

Databases, Cartography, and GI

Geographic and cartographic representation abstract observations and measurements about things and events in the world. These abstractions are only useful if they are saved and stored in formats that enable access to them. Databases provide the most common computerized means to save and store data for swift access. They use different types of computer encoding and organization to access, manage, and analyze this data. Databases follow various principles to create a structure that is extremely flexible for the needs of geographic and cartographic representation; however, the structure is one that at times can be daunting in its complexity. In the information age, databases are fundamental to systematizing representations of the world. At the same time, the use of databases also opens new possibilities for creating many different representations. GI created years ago now can be accessed and combined with other GI.

This chapter is an overview of relational database technologies as they are used in GIS. It introduces the basic principles of the relational database, forms of storage, and applications that perform actions on the data from a database. Data modeling is the term often used to describe these activities.

What Is a Database?

A **database** is a collection of data stored in a structured format using a computer. A database can be thought of as a table, but the distinction is that the table is just one way (of many) to structure the data in the database (see Figure 7.1).

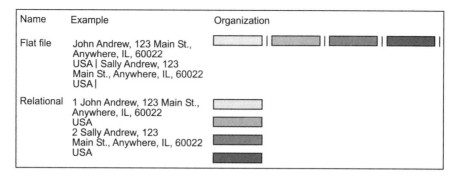

Name	Example	Organization
Flat file	John Andrew, 123 Main St., Anywhere, IL, 60022 USA | Sally Andrew, 123 Main St., Anywhere, IL, 60022 USA |	
Relational	1 John Andrew, 123 Main St., Anywhere, IL, 60022 USA 2 Sally Andrew, 123 Main St., Anywhere, IL, 60022 USA	

FIGURE 7.1. Simplified representation showing the most common types of databases for GI and maps and the important difference between flat-file and relational database types of organization.

The first databases were flat-file databases: computer files of text with one record on each line, usually encoded in the ASCII (American Standard Code for Information Interchange) format, an encoding format still widely used for .txt files. Each entry (e.g., a person's name and address) was separated by a special mark, and commas or tabs separated the characteristics (e.g., name, street, house number, city, postal code, state, and country). Finding a particular person required searching through the entire data file one entry after another. A flat-file database can also be represented as a list, a table, or a spreadsheet. In other types of databases the data is stored as records and fields that correspond to entries and characteristics in the flat-file database. Records and fields have become the accepted terms when working with databases. The term **tuple** is used to represent a single data item in a table. A **field** refers to the division of the data into separate parts of each data item. An **attribute** is the particular entry in a field (e.g., Main Street in the field "Street"). A single database record including all attributes is called an "entity."

The relational database sets itself apart from the flat-file database through the way it stores data and the possibilities for relating data. A relational database stores data in separate files, often called **tables**, which can be related to other "tables" in the database by common fields. A relational database may consist of hundreds or even thousands of tables. Every table in a relational database has a key field that allows each record to be uniquely identified. The key field is usually indexed to speed up operations and is especially important for GI because of the large amounts of data the key field can bring together. In a relational database, attributes can stand in different relationships to attributes in other tables (see Figure 7.2). A one-to-one relationship relates a single record in one table with a single record in another table. A one-to-many relationship relates a single record in one table with multiple records in another table; a many-to-one relationship does the opposite. A many-to-many relationship relates many records from one table to many records from another table. This last relationship is rarely

Relating database tables

FID	Shape*	NAME	STATE_NAME	STATE_FIPS	CNTY_FIPS	FIPS
0	Polygon	Lake of the Woods	Minnesota	27	077	27077
1	Polygon	Kittson	Minnesota	27	069	27069
2	Polygon	Roseau	Minnesota	27	135	27135
3	Polygon	Koochiching	Minnesota	27	071	27071
4	Polygon	St. Louis	Minnesota	27	137	27137
5	Polygon	Marshall	Minnesota	27	089	27089
6	Polygon	Beltrami	Minnesota	27	007	27007
7	Polygon	Lake	Minnesota	27	075	27075
8	Polygon	Polk	Minnesota	27	119	27119
9	Polygon	Pennington	Minnesota	27	113	27113
10	Polygon	Clearwater	Minnesota	27	029	27029
11	Polygon	Red Lake	Minnesota	27	125	27125
12	Polygon	Itasca	Minnesota	27	061	27061

FIPS	POP1990	POP2000	POP90_SQMI
27077	4076	4651	2
27069	5767	5088	5
27135	15026	16076	9
27071	16299	14760	5
27137	198213	193111	29
27089	10993	9937	6
27007	34384	39655	11
27075	10415	10817	5
27119	32498	30353	16
27113	13306	13569	22
27029	8309	8090	8
27125	4525	4155	11
27061	40863	44445	14
27107	7975	7468	9

FIGURE 7.2. Database table relates (or joins) are made by identifying the same attributes in the key fields from two separate tables. In this example, the FIPS code can be used to relate population data to the corresponding counties.

desirable because the meaningful relationships between the records cannot be differentiated from spurious and erroneous relationships. A one-to-many or many-to-one relationship may be called for in a variety of situations (e.g., the cities of one state or the states of one country). These usually reflect a hierarchy or a grouping of attributes and their corresponding records.

The relational database has several advantages for geographic and cartographic representation. First, the conceptual model of the database is distinct from the physical model (how the database is stored and managed on computer hardware). Second, separate tables help maintain the integrity of the potential meaning of database elements. Most relational databases now use structured query language (SQL) for constructing queries involving tables of a single database, or with tables in other databases, even on other computers. Third, the clarity of the relationships aids people using the database with previous experiences of the database. Reliable processing is critical for queries of GI and online maps. Fourth, it is possible to define multiple views of the same data in different database tables (e.g., listing entries by street address or alphabetically by name).

While the relational database is the most common type of database and possibly the only type of database you will ever work with, two other types of databases may be significant. The first of these is a hierarchical database. This database is organized by defining a hierarchy into which all data is stored (e.g., country, state/province, county, municipality). This type of database was frequently used for business transactions, but is being replaced by relational databases. The second is the object-oriented database. In this database, data is stored as objects that not only include characteristics, but also possible actions. The objects in an object-oriented database exist only when the database program is running on a computer (fortunately their characteristics can be stored for later use). Objects act on other objects, receiving and sending messages and processing data. For example, an object-oriented database of sewers may consist of objects with information about the size of the sewer pipe, but also how much water can flow through the pipe in a

IN DEPTH ASCII Characters

The partial list of ASCII (American Standard Code for Information Interchange) characters below contains the characters and their decimal representations, which is how the computer stores them. These numbers are further encoded by the computer in the hexadecimal, and then the binary format. First, you may note that ASCII characters also include nonprintable characters (some are not shown in this extract of the character set). You should also note that letters from other alphabets (e.g., ü or · ą) are missing. ASCII is being replaced by the character set UNICODE, which includes those letters, as well as characters and symbols from many, many other languages. It will be some time, though, before this newer character set is fully phased in.

Dec. Val. Char.											
32	space	48	0	65	A	82	R	99	c	116	t
33	!	49	1	66	B	83	S	100	d	117	u
34	"	50	2	67	C	84	T	101	e	118	v
35	#	51	3	68	D	85	U	102	f	119	w
36	$	52	4	69	E	86	V	103	g	120	x
37	%	53	5	70	F	87	w	104	h	121	y
38	&	54	6	71	G	88	X	105	i	122	z
39	`	55	7	72	H	89	Y	106	j	123	{
40	(	56	8	73	I	90	Z	107	k	124	|
41	)	57	9	74	J	91	[	108	l	125	}
42	*	58	:	75	K	92	\	109	m	126	~
43	+	59	;	76	L	93	]	110	n	127	DEL
44	,	60		77	M	94	^	111	o		
45	-	61	=	78	N	95	_	112	p		
46	.	62		79	O	96	`	113	q		
47	/	63	?	80	P	97	a	114	r		
		64	@	81	Q	98	b	115	s		

Note: The ASCII codes from 0 to 127 are identical to Unicode. More complete ASCII tables are available on the Internet.

minute, and what happens to excess water. Additionally, a sewer pipe may have attributes that can be passed on to other sewer pipe records in the database, or modified to reflect the characteristics of another sewer pipe. Some GIS already use object-oriented databases.

Representing and Communicating

The database's role for storing GI makes it central to the process of communication, also for maps produced from GI. Representation and communication usually involve databases. Databases are part of the technologies we encounter daily and are a field of study, management area, and science in their own right. Most GI and maps only scratch the surface of what databases can be used for, but the two most common uses of databases for GI and maps are as follows:

- Databases store measurements and observations of things and events.
- Databases store the symbols, values, and other graphic elements that help maps communicate (see Figure 7.3).

Graphics drawn automatically by computer software generally are less refined aesthetically compared to graphics humans make directly. Hand-drawn maps and graphics can be used to improve communication. This also becomes necessary if the computer-produced graphic should be revised or GI is unavailable.

The organization of the database tables, records, and fields is called a "data model." The creation of a data model is an important task and needs to be considered in conjunction with the geographic representation, cartographic representation, and communication objectives (see Figure 7.4).

When working with GI or maps, you should be aware of how the database can constrain representing and communicating. This may be the result of using software or hardware that is not adequate to the task, or due to the misuse of the database. The relationships between different database tables can lead to a variety of errors. A common example for roads is that records in one database table use initial capitals and full names for street designations (Road, Street, Lane, Avenue) and another database table uses abbreviations (Rd., St., Ln., Ave.) or fails to use initial capitals. In most cases the database software will not relate the two tables because at the database's level of analysis (e.g., ASCII code) the street designations "Road" and "Rd." are different and unrelated.

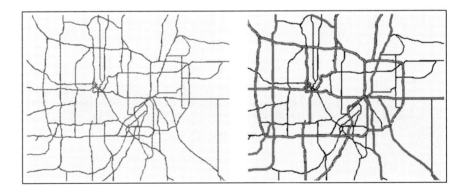

FIGURE 7.3. Highway and major road network in the Twin Cities, Minnesota, United States. The map on the left shows the roads without symbols. Storing or calculating the symbol values from database fields or sensor input help reliably communicate traffic conditions.

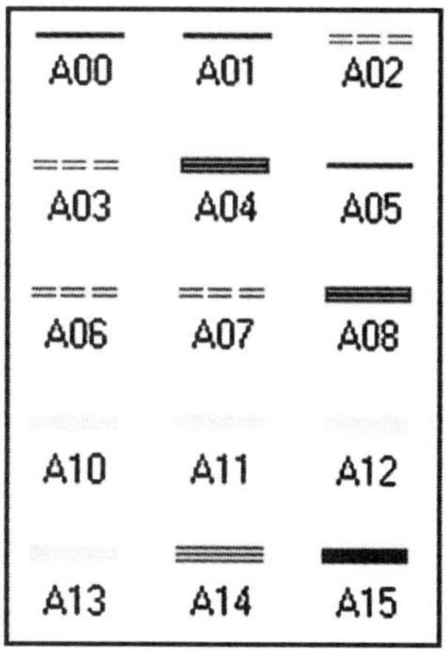

FIGURE 7.4. Examples for symbols used in classifying roads in the United States. The use of conventional symbology and storage of the corresponding data helps to ensure reliable communication with the maps created from this data.

Data Types

Database representation in most relational databases is restricted to a limited number of data formats. The terms used here have other analogous terms, some of which are indicated.

Integer: Whole numbers (1, 2, 3, 4) are stored as integers.

Floating-point: Numbers with decimals (1.1, 2.4, 5.4) are stored as floating-point numbers.

Character string, fixed, or variable length: Also called "text," character strings can be usually stored only in fixed-length fields (see Figure 7.5). In some cases the database software provides variable-length fields. If a word or text is longer than the fixed length of the field, the characters after that place will not be stored.

Date and time, time interval: Because of the unique ways for storing date and time, most database systems offer separate data types for recording date, time, and, in some cases, time interval.

Simple large objects: Any kind of data (including images, word processor files, and spreadsheets) can be stored in a database as a simple large object as binary data. Because this data type is most often used to store binary data, especially images, it is often called binary large objects (BLOBs).

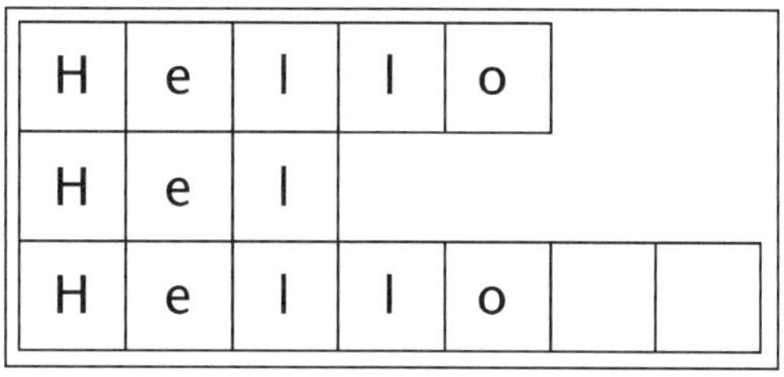

FIGURE 7.5. Examples of proper field length (top); too short field length (middle); too long field length (bottom).

An important issue to consider practically when creating a database is specifying the length of each field, also called "precision." If the field is too long, the database may require a great amount of computer storage space. If it is too short, attributes may be truncated (cut off) possibly making it impossible to know what the attribute actually records.

Data Storage and Applications

Considerations in designing a geographic representation or cartographic representation are the main factors determining how data is stored in a database. The available data types and the allocation of storage for attributes also play important roles. If the GI or map should show demographic characteristics of an area, most of the data will be stored in integer format. Data showing ratios will require floating-point fields. Text-type fields can be added for notation. For an application modeling erosion processes, the data will also be mainly numbers, but the types of observations and analysis will require mainly floating-point data types. Of course, if the geographic representation has led to personal addresses stored in a single field as a character string, it will be very difficult, possibly even impossible, to identify only those people in the database that live on Main Street. Most commercial GIS offer some abilities to control the database storage of attribute data. Vector geometry and raster data also rely on databases, but they are often made inaccessible as part of the proprietary software. Export formats make it possible to use the data in multiple applications.

The application type should guide practical considerations of which data types should be used in analysis and communication. For many purposes, observations and data recorded in numerical formats are the most flexible. They can be transformed and analyzed with other numerical data. Character strings are useful for recording the names and designations of things and events; BLOBs are usually used for images; data and time data types are used to record when things were recorded or events took place.

Entities and Relationships

A key part of working with databases is creating a data model that accurately and correctly shows things or events and their relationships. The clarity of this data model is important for others who need to understand the geographic and cartographic representation, or perhaps just the data model. A data model should describe each entity and the attributes that are associated with the entity's key identifier. Relationships are based on key identifiers and can be between two unique entities, one unique identity to several other entities, or between several entities. Figure 7.6 shows an entity-relationship diagram for typical information a municipal database on residences might hold.

A variety of techniques have been developed to sketch and make schemes showing the data model. Usually these techniques follow the

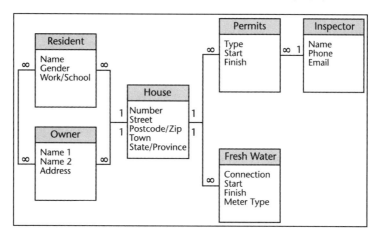

Entity-relationship (E-R) diagram.

entity-relationship conceptual understanding of a relational database. Entity-relationship diagrams, or E-R models, are often made using a graphical form based on the Universal Modeling Language (UML), which offers a systematic way of going from the diagram to conceptual and actual database description (see Figure 7.7).

The capability of creating relations between data is extremely powerful and useful. However, relations work only when data is stored using the same format. For instance, returning to the address example from above, if the entire address is stored as one database field, it will require additional processing to relate this data with address data separated into multiple fields.

If the data can be related, the relation can be permanent or temporary. Any database processing of a relation that produces a single, permanent new table is called a "join"; otherwise it is just a relationship.

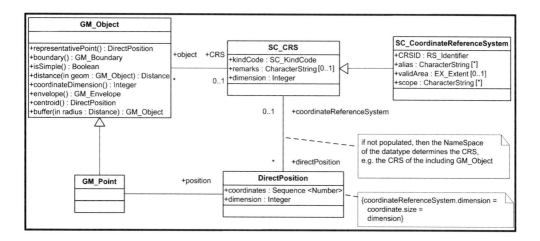

UML diagram for a point object following the ISO UML standard.

Normalization

Database normalization is a process of ensuring that a database can take best advantage of relational database principles and remain accurate and reliable. If you normalize a database you can not only improve its performance, but avoid some organizational and logical errors that could diminish the quality of the database. Data normalization of relationship database technology was first described by Edgar Codd in the 1970s. The first level of data normalization requires that each field contain only one value (e.g., only the house number, not the house number and street name). The second level requires that each value of a record is dependent on the key value of the record (e.g., the name of the person). In the third level, no fields depend on nonkey fields (e.g., a "years at residence" field must be related to the name of the addressed person, not the street number).

Data Modeling, Geographic Representation, and Cartographic Representation

The consideration, inclusion, and representation of the complex spatiotemporal relationships in a database pose a number of challenges that require a thoroughgoing engagement with geographic and cartographic representation. If David Sinton's matrix (see Chapter 2) provides a means of conceptualizing observation, measurement, and storage of the data that describe a single geographic feature or event, the data modeling for a database must consider multiple features and events as well as their relation to each other. In addition, data modeling takes cartographic representation into account in varying degrees. The type of media, the projection, the coordinate system, and symbolization all influence data-modeling decisions. Relational databases have several advantages for flexibility and reliability in addressing these issues. The relationships between tables can reflect different relationships between things and events, and multiple relationships, representations, and types of communication can be part of the data model. The topological modeling of data and relationships of much of the vector data in a GIS supports error detection, address matching, and geocoding. Of course, challenges exist when developing new GI when some GI already exists and when GI from different sources should be combined.

Summary

Databases provide the most common computerized means to save and store data for swift access. This data is based on observations and measurements about things and events in the world. Their design should consider both geographic and cartographic representation issues. Based on the nature of the intended GI uses, the creation of a database could consider only geographic representation. Taking cartographic issues into account is usually very advisable.

GIS databases are in almost all cases relational databases, which have great flexibility for GI and cartographic needs. You can think of a database as a set of tables that can be put into relationship with each other based on characteristics, or what are usually called "attributes." They are stored as one of several data types: integer, floating-point, character, date and time, or simple BLOBs. These tables are different from spreadsheets because each value of a characteristic is kept grouped into a record of all recorded characteristics for that database entity or object. Tables in relational databases are related using the entity-relationship model. These database tables are also used for recording attributes that are used for the symbolization of things and events. Data modeling plays a key part in preparing the geographic and cartographic representation of GI or maps.

REVIEW QUESTIONS

1. What is a database "join" operation?

2. What are the common field types used to store data?

3. What is a flat-file database?

4. What is the relationship between database representation and geographic representation?

5. What is the difference between *logical* and *symbolic* representation?

6. What is tabular information?

7. What is a relational database?

8. What does "database normalization" refer to?

9. What is an entity-relationship diagram?

10. How are cartographic symbols stored in databases?

ANSWERS

1. What is a database "join" operation?

   A "join" operation permanently links two database tables based on a common record value.

2. What are the common field types used to store data?

   Common field types are character, integer, real, binary, exponential, and image.

3. What is a flat-file database?

A flat-file database stores values in rows and attributes by columns. It is one way of representing entities in a database.

4. What is the relationship between database representation and geographic representation?

Database representation is how items are symbolically stored and manipulated in a database. It is based on a geographic representation. For example, a road, geographically represented as two lines, with each line with attributes indicating the number of lanes in a constant direction, can be represented in another database as a single geometric line with two values indicating the number of lanes in each direction.

5. What is the difference between *logical* and *symbolic* representation?

Logical representation is how *symbolic* representation is systematically recorded and stored in a database.

6. What is tabular information?

Tabular information is database data represented in a meaningful tabular form.

7. What is a relational database?

A relational database is a database system that allows for the association of data from different tables.

8. What does "database normalization" refer to?

Normalization is the process of systematizing all relations among database elements and tables for consistent storage and efficient access of data. This reduces data redundancy and improves software and hardware operation.

9. What is an entity-relationship diagram?

A figure to show the conceptual model of a database including all entities and relationships.

10. How are cartographic symbols stored in databases?

Each software package uses its own specific solution, but generally a table of graphic symbols in the software is associated with graphic commands to draw the symbols and related to individual values in a database.

Chapter Readings

Martin, J. (1983). *Managing the data-base environment.* Englewood Cliffs, NJ: Prentice Hall.

Rigaux, P., & Scholl, M., et al. (2002). *Introduction to spatial databases: Applications to GIS.* San Francisco: Morgan Kaufmann.

Shekhar, S., & Chawla, S. (2003). *Spatial databases: A tour.* Upper Saddle River, NJ: Pearson.

Web Resources

⟳ Note: You can find thousands of excellent descriptions of databases on the web. They range from those that are very technically advanced to those that are very simplistic. A search on the words "database introduction" should give you a list of many from which you can quickly choose the introduction best suited to your needs.

⟳ Shashi Shekhar maintains a website with a few chapters online from his book introducing spatial databases (highly recommended, especially for people with some programming or IT experience). See *www.cs.umn.edu/research/ shashi-group/Book*.

⟳ A short introduction well suited to people with only a little computer experience is available at *www.awtrey.com/tutorials/dbeweb/database.php*.

⟳ For an introduction to SQL, try the interactive tutorial at *http://sqlzoo.net/*.

EXERCISES

1. Normalizing a Database

Give students a nonnormalized database of cartographic symbols and features. Have them normalize the database by using a provided symbol table.

2. Create a Theoretical Model of a Geographic Relational Database

Objective

Learn principles of databases, geographic representation, and cartographic representation.

Activities

Students should work from a partial model related to their interests or field of studies to add entities related to a particular application to geographic data.

CHAPTER 8

Surveying, GPS, Digitization

Collecting and communicating reliable GI about things and events requires knowing in a systematic fashion where these things and events occur. Projections, location systems, and coordinate systems provide key geographic reference frameworks for systematically locating observations and measurements of distinct locations. Geographic and cartographic representation rely on the collected location information to make accurate and reliable GI (see Figure 8.1). With various techniques of recording location, surveying, GPS, and digitalization are three generic ways of recording the locations and characteristics of things and events by directly observing them or indirectly measuring their location. In all three forms of location measurement, the collection of positional information requires the systematic collection of measurements.

Geography distinguishes between position and place, though the terms in many other usages are often synonymous. "Position" refers to the systematic measurement of the place associated with a thing or event. "Place" only refers generically to the site, usually referring to something more familiar. For example, the *place* where the Eiffel Tower is located is Paris, France. The *position* of the Eiffel Tower is approximately East 2.37 longitude and North 48.7 latitude. We should also note to use the word "approximate" when referring to position unless we have accurate measurements of location in a geographic reference framework.

Considerations of geographic and cartographic representation issues have strong impacts on the methods and techniques used for collecting positional and attribute data. How we wish to show something by itself and in relation to other things and events involves defining a number of characteristics which in turn specify how the position of objects is determined. For

FIGURE 8.1. Professional surveyors need high-accuracy equipment such as this prism pole, which reflects the laser light used in detailed surveys. The high accuracy of this equipment supports the creation of very accurate and reliable GI.

instance, the location of a forest may be known reliably, but the location of the trees in the forest is another matter. If we need to know the location of the trees we have to decide what a tree is (to avoid including bushes, no matter how large) and where a tree is. For example, these three options for surveying the location of trees could be considered:

1. The tree is located at the center of its trunk at breast height.
2. The tree is located at the northernmost point of the stem, not including surface roots.
3. The tree is located at the center of its canopy, that is, the maximum reach of its branches and leaves.

Depending on our purpose and how we want to geographically and cartographically represent the tree, one of these three approaches or perhaps a different approach would be used. If we are conducting an environmental analysis, knowing tree locations with an accuracy of 1–2 feet or even meters may be sufficient. If we need to be more accurate, as accurate as a fraction of an inch or a few centimeters, we need to rely on the help and services of an expert surveyor, or a geodesist as they are also called.

The discipline of surveying and geodesy specifies methods, techniques, and procedures when high accuracy is required for legal, building, or other purposes. Surveyors are often called upon to meet legal requirements, but they could also be used to satisfy our desire to know as accurately as possible where things are.

Global navigation satellite systems (GNSS), still widely referred to as GPS, are commonly used for recording locations for a variety of applications.

It usually works less well in forests and where there are other obstructions to the signals (for reasons discussed below), but can still be used. If costs are a major issue—for any number of reasons—digitization of existing materials may be a viable option, provided they are available and this form of use is permitted. Many copyrights on maps prohibit using them as the basis for digitization. Assuming that the maps are available and not copyrighted, digitizing GI can be a good compromise and a reasonable way to collect positional information. Open Street Map (OSM) and similar geographic data sets are available under a very broad Open Data Commons Open Database License or similar license that allows generally for free use and reuse of the data as long as users credit the source of the data.

Surveying

With the increasing use of specialized technologies, surveying has become a complex field, but very basic surveying techniques for hobby or curiosity can be practiced by most people. These techniques are elementary and the process requires a minimal amount of mathematics and geometry. The emphasis in this presentation is on the broad understanding of surveying, but this section will also lay out some of the key issues for advanced legally and disciplinary regulated surveying. More and more people survey, which makes it ever more important to know what surveying is and why and when regulations and licenses of surveyors are necessary.

What Is Surveying?

Surveying, broadly understood, is the field collection of positional and attribute information using direct and indirect measurements. More narrowly understood, as a discipline, surveying is the regulated methods, techniques, and procedures of position determination for legally regulated activities, engineering, and other activities requiring certifiable accuracy. Surveying is also known as geodesy in many areas, especially when very recent technologies have become a mainstay of surveying activities.

We can define surveying as the systematic collection of positional location and other location-related characteristics. It is an organized activity using known coordinate systems and procedures for attribute collection based on geographic and cartographic representation. The collection of positional and attribute information in the field must resolve the problems of reducing measurements from the infinitely complex earth to observations that correspond to the geographic and cartographic representation.

Advanced surveying to fulfill the needs of construction and legal requirements is a very specialized discipline. Technologies and methods define the practices of surveying; laws and regulations define the standards and practices.

IN DEPTH Copyright Issues for GI

Almost all GI is protected implicitly or explicitly by copyright, by open-record laws, or by contracts. The only blanket exception is for GI collected by the U.S. federal government. This, however, applies only to civilian agencies. Military agencies (including the Corps of Engineers) are exempt. Individual states have their own laws regulating the use of copyright for their agencies and other government agencies (towns, cities, etc.) in that state. Private companies have copyright on the GI—for example, a map made by a surveyor, unless otherwise defined or regulated.

Copyright sets out to motivate the expression of ideas by restricting how original works in a tangible medium may be used by someone else. In GIS, the work being protected is the geographic or cartographic representation.

Charging for GI is commonplace all over the world. Copyright is a legal way to ensure that people who use GI compensate the creators for their work. When a person uses copyrighted material, he or she has to request permission and/or reimburse the owner. This can get very expensive for the purchaser and very lucrative for the seller, so there are many people struggling over copyright and seeking exemptions that allow them to do what they would like to do without compensating the owner. The U.S. government took the stance over 200 years ago that copyrighting material created for and by the government would hinder commerce and be an imposition for the development of the economy. In considering the use of GI in the world, the results are clear: the U.S. GI economic sector is vibrant and there is widespread (even global) access to U.S. federal government GI. U.S. state governments often take a different view enforced through open-record laws. Some allow free access, some charge. Usually the laws and regulations offer a variety of exceptions, but some states have decided that because GI costs money to produce, users should be charged.

The laws of states regulating access to GI include open records laws, which are related to the Freedom of Information Act. In the United States these laws regulate access to GI (and other types of information created by the government). In Europe the Freedom of Information Act has led to a number of attempts to acquire GI from government agencies. The INSPIRE project is increasing the availability of GI. This is still a cumbersome process and may not lead to the desired results if parallels to the open records laws can be made. For example, digital access in an open-record law may be sufficient (legally) if people can come to a government office and sit down at a computer and access the data. This is access, but not necessarily the access to the original GI on a different computer that most people might expect. In this, and similar ways copyright often becomes a way to protect resources rather than to motivate the creation and dissemination of ideas.

Brief History of Surveying

Even without telescopes, tape measures, or lasers, ancient surveyors could do work of astonishing accuracy. The pyramids in ancient Egypt are evidence of that accuracy that exemplifies the advancement of Egyptian surveying. Even older map fragments found in Mesopotamia (modern Iraq) point to that society's advanced surveying techniques. Even if we can only puzzle over the construction of neolithic monuments in Stonehenge, Easter Island, and other places, these monumental works' locational accuracy in reference to movements of the solar system demonstrates their creators' surveying skills.

For most of known history, surveying has been a very stable discipline, only changing as new instrument-making technology advanced and survey accuracy increased. If a surveyor from ancient Egypt had been able to travel through time and go to any Western country up until the 1880s, he or she would have found the accuracy of survey measurements greatly advanced, but the techniques and basic instruments remarkably similar. Surveyors used a chain of fixed length as a common instrument to measure distances for surveys in many parts of the United States until the 20th century.

During these four millennia, surveying involved numerous techniques that can be simplified first into distance and angle measures and second into leveling. Distance and angle measure involved the use of a plane table to make situation drawings "in the field" and devices such as telescopes to make accurate measures (see Figure 8.2). Leveling was done with plumb bobs and water and mercury levels to accurately measure changes in elevation between locations and their respective heights. Very accurate surveys were conducted to create a national geodetic control network that could

FIGURE 8.2. Surveyors at work with a plane table, in the past a common way to help make accurate surveys.

FIGURE 8.3. Geodetic markers are part of national triangulation networks and are connected to geodetic datums.

serve as the backbone for additional accurate surveys (see Figure 8.3) connected by this network. Surveying was often connected to navigation (Figure 8.4), and most surveyors were capable of navigation by sextant using the stars or sun.

Basic Field Survey Techniques

The most elementary techniques for collecting positional information only require instruments for measuring distances and angles. Usually collected by making drawings on a plane table (imagine a big flat breadboard kept level on a tripod), a basic field survey starts at a point with a known position. Measurements taken for a survey continue through a series of distance and angle measurements which are verified against each other using trigonometric equations. At the same time changes in elevation are recorded. Figure

FIGURE 8.4. A compass can be used for navigation and surveying.

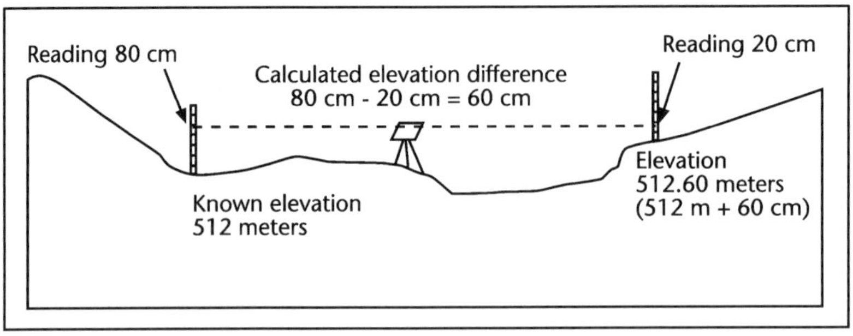

FIGURE 8.5. Basic concepts of leveling for a survey. See the text for further discussion.

8.5 provides a schematic illustration of the process with a plane table and two poles for recording the elevation differences. Trigonometric equations would be used to verify the measurements and to assess their accuracy.

A survey of positions collected in this manner may be accurate by itself, but it could not easily be combined with other surveys and other GI to make maps. Lacking a clear relationship of at least one point (four are for statistical reasons the practical minimum to consider) to a vertical and horizontal datum, it would be very hard to connect the surveyed positions to any coordinate system. Use of a geodetic control network helps assure that multiple surveys can be combined.

More advanced survey techniques rely on defined procedures and rules. These techniques are a basic part of a trained surveyor's skills because of much greater error control and accuracy measurements than possible with basic field survey techniques.

GPS and GNSS

Many people have heard of **GPS**, or the global positioning system, many more have used it without even knowing it. A common feature in new cell phones, cars, and boats, GPS has been used for years by a wide and varied group of users including trucking companies, buses, hikers, taxis, surveyors, and airlines. GPS has become commonplace because of its ease of use and accuracy in determining location. While its level of accuracy is insufficient for professional surveyors, for most people, most of the time, GPS is accurate enough. Car navigation systems, which are becoming very common, offer the ability to show where the car currently is and to get instructions about how to get to another place. The instructions can be shown on a display with a map or spoken, making it easier for the driver to remain concentrated on his or her driving. The navigation systems generally work well, but can become a bit nagging when the computerized voice incessantly rattles off the changing names of streets and twists in a road. Other applications—for example, navigation systems for visually impaired people—point to the many potentials of GPS.

Because of its ubiquity, importance for so many different activities, and the creation of other satellite-based systems for positioning, the term GPS, which refers to only the U.S.-funded satellite-based position-finding system, is slowly being replaced by the term "global navigation satellite system" (**GNSS**), which is the broader term. Because the only other GNSS are currently under development (Galileo) or of use by professionals largely (GLONASS), this book uses the term GPS, although other systems will become more operational. The term GNSS is certainly finding wide usage. For example, although behind schedule, the European Union is currently developing a system, fully compatible with GPS, called "Galileo." Two prototype satellites were launched in late 2005 for testing. The full system, with 30 satellites and offering better coverage of polar regions, should be operational in a few years. The People's Republic of China is developing the second generation of another system, called BeiDou Satellite Navigation System-2, previously called COMPASS, and scheduled to be fully operational by 2020. It is already in use in Asia with reported accuracy of 10 meters.

What Is GPS?

The global positioning system (GPS) is a system of satellites launched and maintained by the U.S. Department of Defense (see Figure 8.6). Over 50 satellites have been launched, each about the size of a school bus (see Figure 8.7). The system costs around $400 million yearly to maintain, but is freely available all around the world. The accuracy and availability can be limited by the Department of Defense if they see a need through selective availability, which degrades the signals received by commercial GPS units.

With a GPS receiver, which can be a computer chip close to the size of a postage stamp attached to an antenna, a device can receive and process the GPS satellite signals and determine location and elevation. How this works is quite complex, but the general idea is rather straightforward.

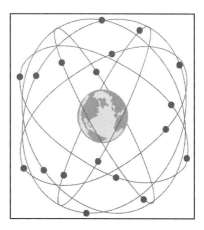

FIGURE 8.6. Idealized drawing of GPS satellites in orbit.

FIGURE 8.7. The only GPS satellite on public display.

Instead of using a measurement of distance, as in surveying, GPS uses the time it takes radio signals to travel. With much simplification we can say that the GPS receiver calculates the difference between its own clock's time and the time communicated in signals from GPS satellites, then uses this difference to calculate the distance between the receiver and the satellites (see Figure 8.8). The time difference is detected in the difference between the signal sequence (a binary signal called "pseudorandom") received by the GPS receiver and the signal sequence it has. Each satellite broadcasts a

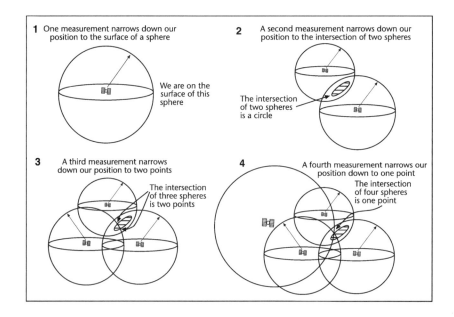

FIGURE 8.8. Determining position with GPS satellites and a ground-based receiver.

signal that contains data about the satellite and the time on its clock. The GPS receiver's time calculations also should take a variety of interferences into account, especially interference in the atmosphere of the earth that can slow down the transmission of a signal from a satellite. If the GPS receiver is traveling, the corresponding movement must also be taken into account.

Obviously some limits to the accuracy of the GPS measurements arise, related to the number of satellites available. The most complex part of GPS positioning is the determination of location. With the signal from one GPS satellite, a GPS receiver can only determine how far it is from that satellite, but not where. It could be anywhere on an imaginary sphere drawn at that distance from the satellite. To determine the position of the GPS receiver, signals from at least four GPS satellites are needed if no other information is available. If the elevation of the GPS receiver is known, only two satellites are needed to determine position.

With the variability of the atmosphere, the movement of the GPS receiver, and possible obstructions in the local environment of the GPS receiver, the accuracy of GPS positioning may be limited. The factors that reduce the accuracy are summarized in the measurement called the positional dilution of precision (**PDOP**). Larger values indicate less accurate GPS positioning. If the values are greater than 8, then the positional location provided by the receiver is very inaccurate. Values less than 4 are a good indication of high accuracy.

Various other factors impact the positional accuracy of GPS measures. The accuracy of most GPS receivers is less than 3 m under ideal conditions. The accuracy is often even around 1 m. If the PDOP value is less than 4, the positional values may even be accurate down to 1 or 2 m. Most GPS receivers take atmospheric interferences into account by using information about the atmosphere at a given time. More accurate receivers compare the speeds of the GPS satellite signals to calculate the reduction in positional accuracy. Most receivers also take the reflection of GPS satellite signals off of the ground, vegetation, and buildings (**multipath error**) and changes in satellite orbits into account. A receiver can reduce error by choosing satellite signals based on the characteristics that introduce error into the position locations and using satellites that help produce the mathematically most accurate results. The quality of the receiver also contributes to accuracy. Higher quality receivers have more or better procedures for reducing errors.

Greater accuracy can also be achieved through a variety of techniques, regardless of the cost of the receiver. The most common of these (and the most widespread) is called the differential global positioning system (**DGPS**; see Plate 7). DGPS relies on a fixed GPS receiver that can calculate GPS changes in its position against the accurate surveyed position. The difference indicates the current amount of error in the GPS system for a particular area. The fixed-location DGPS receiver broadcasts a correction signal used by GPS receivers to adjust the GPS signals and get highly accurate (down to centimeter or inch accuracy) positions. The most common DGPS is called the wide area augmentation system (**WAAS**) and is available in North America. Similar systems, but with different names, are available in many other parts of the world.

Applications

Most people who have used a GPS probably can't imagine any limit to its applications. Even if its shortcomings are grievous (it can't be used indoors, nor very well in a forest, or where there are many tall buildings or cliffs), solutions have been developed to these problems. Usually these solutions involve broadcasting radio signals or pseudo-GPS signals that are highly accurate. The configuration of these systems is very complicated and requires large institutional investments. Most are made by governments. For instance, the European Union is developing a high-accuracy network (along the lines of the U.S. WAAS) for navigation purposes. Even if a GPS receiver lacks the ability to use these extra networks, GPS can still be used in a number of applications, some of which are described here.

Vehicle Navigation Systems. More and more people use GPS-based systems in cars; many more have benefited from the use of GPS in buses, trains, and trucks (see Figure 8.9). The GPS receiver may be hidden, but may be critical for the taxi company to find out which taxi is closest to you when you call for a pickup. A GPS receiver can help a trucking company better organize deliveries to minimize the fuel used. A bus may have a GPS installed to help the bus company indicate to passengers how long they need to wait for the next one.

Navigation systems are used for more than vehicles on land. They are also widely used for nautical and aeronautical navigation. They have become for many sailors irreplaceable because they work regardless of the weather and can easily be combined with computerized chart information. Almost all planes use, or will use, GPS. Together with high-precision positional transmission, planes can use GPS-based systems to land in any weather with centimeter precision.

Hiking. More and more hikers turn to GPS to help them find out more exactly where they are and to help them to plan a route before they go. GPS may not be reliable in canyons or along steep cliffs, but in most situations and weather it provides accurate positional information. Some mapmakers

FIGURE 8.9. GPS navigation system built in as an integral part of a car's dashboard. GPS has become a widely used and nearly ubiquitous technology in navigation systems, mobile phones, and other consumer and professional devices.

have started to change their map designs to make it easier for hikers to use. Some tourist areas offer GPS for people to help them follow a certain tour.

Aids for the Visually Impaired. Combined with acoustic or tactile signaling devices, GPS can be used to help visually impaired people find their way in new settings and navigate places that rapidly change—for example, a state fair or a college campus, as was done by the late Professor Reg Golledge and others at the University of California at Santa Barbara.

Digitization

When maps exist, it is possible to convert them to GI by using either tablet digitization, heads-up digitization, or a scanner. The reasons for digitizing from maps cover a gamut: the maps may be old and show something that people want to compare to recently collected GI, the maps may be unusual or handmade, or the maps may be the only way of getting the desired GI.

Tablet digitizing involves the affixing of the source material (maps, drawings, etc.) that are georeferenced to coordinates on a table digitizer, a board of variable size. These coordinates are connected to clearly distinguishable features on the map. Survey control points are perhaps most desirable, but if not available using very distinct corners or points is valuable. In those cases, determining the coordinates of the feature on the ground can be a limiting factor. For practical reasons often less suited features have to be chosen. It is possible to use more than four registration points, which can help improve accuracy, or help reveal the relatively low accuracy registration points.

Map features can be digitized in several ways. Most commonly, the location of the digitizer puck, the mechanical pointer calibrated to the digitizer and freely movable, is recorded as different buttons are pressed or as a constant stream of locations. Software translates the location and button values. Heads-up digitizing is similar, but requires that digitized source material be georeferenced to a coordinate system. The material is displayed on a screen and the person doing the digitizing uses a mouse or similar pointer device and presses buttons to record locations. The scanning of existing map material is also common, and because it is mostly automated is very fast compared to tablet digitizing, especially when a large number of maps are to be scanned. But scanning usually requires complex postscanning cleanup. Hand digitizing is generally cheaper, but also generally less accurate. Scanning is expensive for just a few maps, and may be complex to configure, but it gets cheaper if you have a number of maps prepared in the same way. Either way, edge-matching is a necessity because of differences in the way the maps were prepared and errors in entering the data. At times rubbersheeting can be used to remove small discrepancies due to distortions in maps, but trying to remove any larger errors often results in other errors.

Keep the following issues in mind when preparing or working with surveyed, GPS'd, or digitized data:

Accuracy and Precision

The data collection procedures, tools, and techniques should assure the highest level of fidelity to the geographic representation and the cartographic representation possible or feasible. If the accuracy of the map materials is known, you have a great assistance in knowing how accurate the GI is. If not, it becomes very complex and a source of troubles. A rule of thumb is to always be more cautious than necessary when determining accuracy.

Of course, you have to be sure not to mix up accuracy and precision. "Accuracy" refers to the agreement between the GI or map position and the ground position, whereas "precision" refers to the number of digits used to indicate the position. High precision is meaningless for most purposes without corresponding accuracy.

Choice of Positional Collection Technology and Approach

The purpose of and means available for data collection largely will determine the collection technology. The most important additional factor here is often cost. If the data collection has to choose between two methods, generally the lower-cost option will win. The exception would be if the higher-cost option offers additional information, accuracy, or reliability. Of course, the lowest-cost option can easily end up being the most costly in the end. Far too often, people collect GI without thinking through the geographic and cartographic representation.

Closure of Areas and Connections between Lines

A complicated issue for digitization can be making sure that areas are closed (e.g., counties, states, or countries) or lines are connected (e.g., highways, bus routes, or subway lines). This can be remedied by using a tolerance that moves digitized points together if they fall within a specified distance of each other. The tolerance is usually based on the accuracy of the GI. This can be difficult because the proximity of features changes across a map. Buildings are closer together in urban areas, some areas have long and narrow fields, others have very large rectangular fields. The tolerance for connecting points when digitizing needs to be adjusted to the circumstances.

Generalization Effects

When digitizing maps, you need to bear in mind that generalization operations may have moved features on the map to make the map easier to read. This is common in small-scale maps, but also occurs in large-scale maps. If it is impossible to find out how features have been generalized, you can at least use the indicated accuracy of the map as an indicator of how much an individual feature could have been moved.

In examining positional collection technology options, you also need to consider remote sensing, discussed in Chapter 9.

Summary

The GI for maps and other communication should be collected to fulfill requirements arising from issues related to the geographic and cartographic representation. Accurately and reliably locating observations and measurements about things and events requires careful consideration of the options for collection and the issues each option faces. The three generic options are surveying, GPS, and digitization. Traditionally, surveying was the discipline called on for accurate and reliable measurements of position. GPS, which is widely available today and becoming more commonplace, is altering that somewhat, but surveying remains the discipline called on for accurate and reliable location measurements, especially when legal dimensions of the things or events are important. Existing materials can be digitized or scanned, but copyright regulations and limitations should be carefully considered.

REVIEW QUESTIONS

1. What specific steps does the systematic collection of positional location entail?

2. Under what circumstances is surveying legally regulated?

3. What instruments are commonly used today for surveying?

4. What instruments were used traditionally for surveying?

5. What legal issue must be considered before using existing GI?

6. Where is the use of GPS less accurate?

7. Which accuracies does GPS support?

8. How should geographic and cartographic representation be taken into account for data collection?

9. How can generalized GI and maps affect positional accuracy?

10. What are common sources for existing GI?

ANSWERS

1. What does the systematic collection of positional location involve?

   The systematic collection of positional location and other attributes is an organized activity using known coordinate systems and procedures for attribute collection based on geographic and cartographic representation.

2. Under what circumstances is surveying legally regulated?

Generally, surveying is legally regulated when it involves the collection of location information used for purposes of or activities with possible immediate public safety consequences.

3. What instruments are commonly used today for surveying?

Total stations, GPS (GNSS), and laser range finders are among the most common.

4. What instruments were used traditionally for surveying?

The theodelite, measuring tapes, rods, and plane tables were traditionally used.

5. Can existing GI be used in any way?

The copyright status and distribution rights must be assessed before using GI from other sources.

6. Where is the use of GPS less accurate?

Generally, GPS is less accurate under tree foliage, near trees, near cliffs, or near high buildings, all of which can obstruct the GPS satellite signals.

7. Which accuracies does GPS support?

GPS (or other GNSS) is suited for any activities where information about location at a modest accuracy (3–5 m) is needed. For higher accuracies, additional procedures, tools, and techniques can be used.

8. How should geographic and cartographic representation be taken into account for data collection?

The data collection procedures, tools, and techniques should assure the highest level of fidelity to the geographic representation and the cartographic representation.

9. How can generalized GI and maps affect positional accuracy?

Generalization distorts GI and maps in a variety of ways. This distortion reduces positional accuracy.

10. What are common sources for existing GI?

Many government agencies, national mapping agencies, and private companies are potential sources of existing GI.

Chapter Readings

Campbell, J. B. (2011). *Introduction to remote sensing* (5th ed.). New York: Guilford Press.

The mathematical basis for geographic surveying is covered in

Cotter, C. H. (1966). *The astronomical and mathematical foundations of geography*. New York: Elsevier.

For an older, but lucidly presented text, see
Hinks, A. R. (1947). *Maps and survey*. Cambridge, UK: Cambridge University Press.
For a more recent text for students starting with geographic fieldwork, see
Lounsbury, J. F., & Aldrich, F. T. (1986). *Introduction to geographic field methods and techniques*. Columbus, OH: Merrill.
For a history of surveying and map making, see
Wilford, J. N. (2001). *The mapmakers*. New York: Knopf.

Web Resources

- A good glossary of GPS and GNSS terms is available online at *www.gmat. unsw.edu.au/snap/gps/glossary_a-c.htm*.

- More on the surveying of the United States is available online at *www. measuringamerica.com/home.php*.

- An excellent starting point for resources on the history of surveying is available online at *www.fig.net/hsm*.

- An overview of geodesy and its different uses is available online at *www.ngs. noaa.gov/PUBS_LIB/Geodesy4Layman/toc.htm*.

- A very thorough overview of the GPS system focusing on technical aspects is available online at *http://cn.wikipedia.org/wiki/GPS*.

- Trimble, a large GPS hardware and software provider, offers an animated GPS tutorial that is available in their Learning Center online at *www.trimble.com/ tr_open_main_courses_asp*.

- An overview and details about the roles of GPS for aeronautics (and beyond aviation) is available online at *http://gps.faa.gov/index.htm*.

- For information related to European GNSS activities, see *www.esa.int/esaNA/ index.html*.

- The American Congress of Surveying and Mapping (ACSM) offers much information about current surveying training and activities online at *www.acsm. net*.

EXERCISE

GPS and Navigation: Good for What?

What are the advantages of using GPS for navigation? What kinds of navigation benefit the most? Why? What are some of the disadvantages? Are there potentials to abuse GPS and collect personal information about people?

Finally, what about the use of GPS for surveying? Do you think any person should be allowed to use GPS for surveying?

EXTENDED EXERCISES

1. Basic Surveying

Objective

Learn basic concepts of surveying, especially triangulation.

Overview

Most people don't think about it, but without surveyors not much would happen in a modern world. Houses, banks, roads, bridges, airports—all structures—rely on accurate surveying. Property ownership is also surveyed, and when it's not done properly expensive legal conflicts are usually unavoidable. The mathematical foundation of surveying is Euclidean geometry. This is the oldest geometry in the Western world and the one that approximates very well our actual experience of distance relationships.

In this exercise, you will learn a little about one of the oldest and most reliable surveying techniques, the *plane table survey*, by preparing a simple survey and doing some geometric evaluation. While the results of this exercise will not be accurate, the technique you will learn, when conducted with the appropriate instruments and robust procedures, will allow you to survey many things. And you will have a new insight into why mathematics is important for maps.

Concepts

Many technologies are used in surveying. The plane table survey is an incredibly simple technology for fairly accurate surveys. The technology usually uses a large 2' × 2' board positioned over reference points. By using an accurate instrument for sighting points (one type is called an "alidade") and by keeping a consistent scale for measuring distances, a surveyor measures angles that accurately bisect position points. In other words, a traditional surveyor uses angles to determine distances and rarely measures distances directly. (Using computer-based surveying devices employing laser range finders, surveyors now are likely to measure distances directly.)

For this exercise, you will be using a simple piece of paper as a pseudo-plane table working indoors (making it possible to do this exercise any time of year). In comparison to working with a plane table, this diminishes the accuracy of the survey; however, the concepts and techniques remain the same. Using measurements of angles, you will be applying Euclidean geometry's *law of sines* to construct a locational survey of items in the classroom. Make sure to look at the law of sines example before starting with the exercise.

Exercise Steps and Questions

Preparation

You need to have a piece of paper (called a worksheet) for drawing your measurements and making the basic calculations, which you should place on a pad or spiral notebook. You will also need a straightedge ruler and a few colored pencils. Your instructor will provide you with a protractor that you will use to measure angles.

Your instructor will have created several baselines in the lab room. Each line is measured in centimeters. Each baseline forms a side of the triangles you will construct to locate objects in the room.

What to Do

First, form a group of four or five people. Prepare your lab instruments and get one protractor for each group. *Each person should complete a sheet indicating all measures, constructions, and calculated distances.*

You will be surveying and determining the angles from the baselines to *five objects* in the room. Example objects are:

- Thermostat
- Window levers
- Wall sprinklers
- Emergency lights
- Clock
- Light switches

Step 1

To start with your survey, go to one of the baselines. Two points are indicated, one labeled *A* and the other *C*. Draw a line to scale on your worksheet positioned to fit the surveyed elements and label the points. If your baseline is in the front of the room, put it at the bottom of the page. No matter where you are in the room, remember to always keep the orientation of your page. To figure the scale, set up a ratio between map units and ground units—for example, 1 inch on paper = 100 cm in the room, or 1 cm on paper = 25 cm in the room.

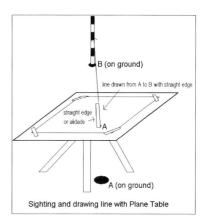

Sighting and drawing line with Plane Table

Step 2

Put point *A* on your worksheet directly over the corresponding point *A* on a baseline in the room. *Accuracy here is very important.* Make sure to keep the paper stable after you have found the right position. Now, take your straightedge and point it from the point on your worksheet to the object you will survey—clock, thermostat, or the like. Make sure you are very accurate in drawing the line with one of your lighter-colored color pencils. Draw a straight line at least long enough to cover the distance in scale (you will get better at this with experience). Move to the second point (point C), reposition your paper so that point C on the paper coincides with point C on the ground, and repeat the sighting with your straightedge and drawing of a line. The two lines should meet, forming a triangle. Help the next person set up, checking to see if he or she is also following the correct procedure.

Step 3

Now, use the protractor to determine the angles of the triangle you have just drawn. The protractor has two degree indications. Negative angles run from left to right, positive angles run from right to left. Use the indications that correspond to the direction of the angle, or direction of the "base" of the angle—for example, if the base of the angle (also called the "initial side") points to the right, use the angle indicators that run from right to left. Write the angle measurements on your worksheet together with the figures.

You can now use the law of sines to determine the distances from the baseline to the surveyed object. You only need to calculate one distance, but calculating both distances will be helpful.

Repeat Steps 1, 2, and 3 for the other four objects, making sure to position your worksheet accurately and measure angles very carefully. Put all your measures and the results of your calculations down on your worksheet. Work together with other people in your group to make sure everybody has the same (or almost the same) measures.

Evaluation

When you have finished surveying and calculating distances, answer the following questions.

Question 1

Draw a line in another color connecting your surveyed objects on your worksheet. Does it look like a straight line approximating the wall? Compare your measurements and calculations for each of the five objects you surveyed. How accurate were your measurements and calculations? What is the difference between your calculated positions and measurements? What explains the difference?

Question 2

You surveyed in only two dimensions. Would adding a third dimension for height make your survey less accurate or more accurate? Why? What about for more precise surveying work in general? What is the name of the process a surveyor conducts to assure accurate height measurements?

2. The Law of Sines and Euclidean Geometry

Introduction

In this exercise, you will be using Euclidean geometry, named after the ancient Greek mathematician Euclid who lived around 300 B.C. and who wrote 13 books about mathematics collectively called *Euclid's Elements*. It is the most established approach to codify perceptions of space and motion. Euclidean geometry is also called "classical geometry" because many other people contributed to it and added to it over the centuries. Euclid's geometry consists of 10 axioms for fundamental geometrical relationships, such as the sum of the angles in a triangle always equals 180 degrees.

Even though Euclidean geometry is very old and physics has modified its applicability to certain phenomena that are better explained by Einstein's theory of relativity, quantum dynamics, and so on, Euclidean geometry is very important for many modern activities ranging from surveying to computer-aided design, computer vision, and robotics. If you have ever played, or seen, a new videogame and been amazed by the graphics, a large proportion of the math behind those graphics is based on Euclidean geometry.

The law of sines is one of the most fundamental parts of Euclidean geometry used by surveyors. It expresses the relationship between an angle and its opposite side. In right angle triangles, the sine is the relationship between the opposite side and the hypotenuse. In any triangle, the ratio of one side to its opposite angle is the same as the ratio of any other side to its opposite angle. Expressed mathematically:

$$\frac{a}{Sin\ A} = \frac{b}{Sin\ B} = \frac{c}{Sin\ C}$$

The law of sines is related to the law of cosines and also to the law of tangents. These are more complicated formulas for solving for the lengths of sides and size of unknown angles.

Using the Law of Sines in Surveying

The law of sines is used to solve the length of an unknown side when you know the length of one side and two angles. In this example, I go through the steps to find out the length of *c* in this figure.

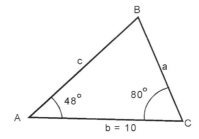

In the law of sines, all ratios are equivalent. If we know any three terms from two ratios, we can use basic algebra to solve for the unknown term. In this case:

$$\frac{a}{Sin\ A} = \frac{b}{Sin\ B} = \frac{c}{Sin\ C}$$

Now substituting the known terms

$$\frac{b}{Sin\ B} = \frac{c}{Sin\ C}$$

then using a sine table or the first three digits of a calculator's sine:

$$\frac{10}{.788} = \frac{c}{.985}$$

finally solving for *c* by multiplying both sides by .985

$$c = \frac{10(.985)}{.788} = 12.50$$

Additional Resources

If you want to find more information about the law of sines or Euclidean geometry, see Euclid's *Elements* (with interactive demonstrations) online at *http://aleph0. clarku.edu/~djoyce/java/elements/elements.html.*

Also see Geometry Reference Materials, available online at *http://mathforum. org/geometry/geom.ref.html.*

CHAPTER 9

Remote Sensing

Remote sensing is the collection of data without directly measuring the object. It relies on the reflectance of natural or emitted electromagnetic radiation (**EMR**). EMR can be emitted by the sun (natural EMR) and, for example, sensed by photographic film, or it can be sent by a transmitter and the returned energy sensed, for example, by radar (a type of emitted EMR). Remote sensing has become a key means of data collection for a number of reasons, but mainly because it allows for the systematic and accurate collection of GI.

Remote sensing is defined very broadly in this chapter as a measurement of an object's characteristics from a distance using reflected or emitted electromagnetic energy. This definition means that remote sensing includes all kinds of photography, aerial imagery, satellite sensors, and any kind of laser measurement. Remote sensing involves different types of sensor technologies ranging from photographic emulsions to digital chips (see Figure 9.1). It also involves a vast array of storage media including everything from photographic film to computer files. As you can imagine, data from remote sensing and the sensor technologies themselves are a resource that can enhance the work being done in a number of other fields of study. For example, the discipline of surveying has changed enormously with the introduction of laser-based distance-finding technology.

The reason for defining remote sensing so broadly is that it is a very important GI technology. Remote sensing offers three advantages over other forms of data collection and GI. First, it makes it much easier to systematically recognize things and events over a large area. Second, it makes it easier and less costly to revise most maps. Third, digital remote sensing images can be used directly by other applications. There are some caveats to these advantages that you will find out about in this chapter. This chapter is purely introductory in nature and will skim over many of the crucial details and

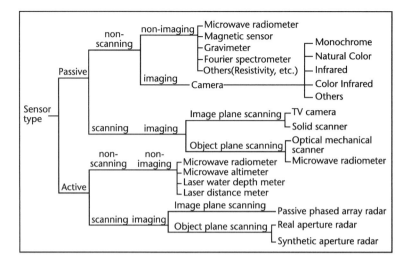

FIGURE 9.1. Different sensor types. A key distinction is made between passive sensors and active sensors. Passive sensors use only reflected electromagnetic radiation (EMR). Active sensors use emitted EMR.

physics, but you should end up with a solid understanding of what remote sensing involves and what some of the key issues and applications for remote sensing are.

Electromagnetic Radiation

Any understanding of remote sensing, regardless of the sensor technology, storage media, or application, starts with understanding EMR. First off, remote sensing's detection of EMR has three characteristics:

1. It generally only detects EMR from the surface of an object, although some sensors allow for penetration.
2. There is no contact between the sensor and the object.
3. All remote sensing measurements use reflected energy (usually from the sun) or emitted energy (e.g., from a radar station or plants).

The EMR detected by remote sensing technologies varies. It depends on the desired application as well as on the cost of different remote sensing data collections.

Spectral Signature

Emitted or reflected EMR varies (see Figure 9.2). These differences are the basis for distinguishing things and events. The reflections and EMR

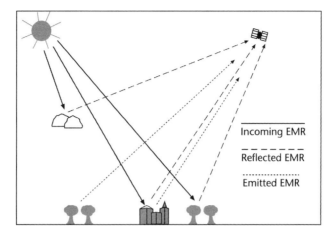

FIGURE 9.2. Emitted and reflected electromagnetic energy that remote sensing sensors receive to create remote sensing data or images.

emissions of a particular thing or event can be associated with a particular spectral signature that is used to identify where these things and events are located in a remote sensing image.

EMR also varies by time of day, season, weather conditions, moisture levels in the soil, wind, and a number of other factors. The physics involved in addressing these variations in emitted or reflected EMR is critical to the success of remote sensing and provides a commonplace solution (see Figure 9.3). This solution, called "**ground truthing**," involves having some people in the field before, during, or after data collection who may take similar sensor measurements or observations. These measurements and observations

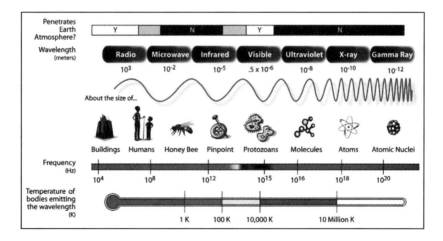

FIGURE 9.3. The electromagnet spectrum showing common examples and corresponding sizes to compare wavelengths.

can be used later to verify the remote sensing image or data and possibly to define correction parameters for adjusting the remotely sensed data to correspond to ground observations (see Figure 9.4). Needless to say, this is highly complex and requires very well trained specialists to assess these factors and detect patterns in the remote sensing data.

Bands

The detection of patterns is helped by the use of different ranges, or bands, of EMR in sensing technology. Each **band**, as they are commonly called, refers to a particular range of wavelength for that sensor. The bands available for a particular sensor depend greatly on the purpose of the sensor and the technical characteristics of the sensor. Some sensors have only a few bands in a narrow range of the total EMR, others are much broader. For example, Landsat 7 has seven bands:

Band 1: 0.45–0.52 µm Blue–Green
Band 2: 0.52–0.60 µm Green
Band 3: 0.63–0.69 µm Red
Band 4: 0.76–0.90 µm Near IR
Band 5: 1.55–1.75 µm Mid-IR
Band 6: 10.40–12.50 µm Thermal IR
Band 7: 2.08–2.35 µm Mid-IR

Figure 9.5 shows the different bands and how they can be combined for an application.

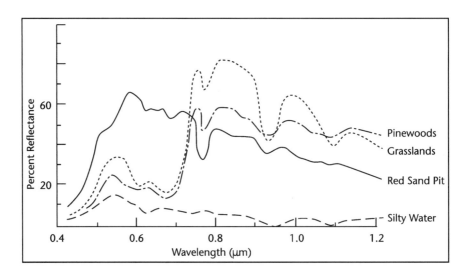

FIGURE 9.4. Examples of spectral signatures and wavelengths of reflected electromagnetic radiation. Note that a micrometer is one millionth of a meter.

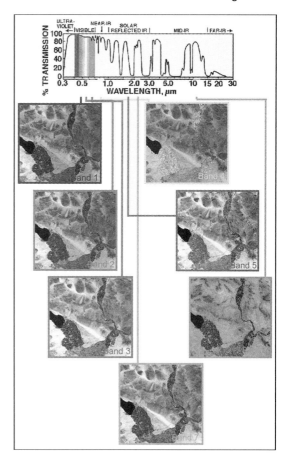

Another widely used satellite, SPOT 5, offers a different set of band-widths. The newest Landsat satellite, Landsat 8, with 11 bands, extends the bands of Landsat 7.

Resolution

Remote sensing distinguishes between spatial, temporal, and spectral resolution. (See Tables 9.1 and 9.2.) **Spatial resolution** is the size of the unit recognized by the sensor, **temporal resolution** has to do with how often a satellite passes over and/or takes readings of the same spot, and **spectral resolution** measures the range of wavelengths the sensor can record. A raster cell is often also referred to as a "pixel."

Spatial resolution is usually given in a distance measurement (see Figure 9.6). For example, most SPOT sensors have a resolution of 10 m; some have a higher resolution of 2.5 m. The bands of Landsat 8 generally record in a resolution of 30 m, but one band collects panchromatic data at 15-meter

TABLE 9.1. SPOT 5 EMR Spectrums and Bands

Electromagnetic spectrum	Spectral bands
Panchromatic	0.48–0.71 μm
Green	0.50–0.59 μm
Red	0.61–0.68 μm
Near infrared	0.78–0.89 μm
Midinfrared (MIR)	1.58–1.75 μm

TABLE 9.2. Comparison of TM and ETM+ Spectral Bandwidths for Landsat 5–TM and Landsat 7 (Source: *http://landsat.gsfc.nasa.gov/guides/LANDSAT-7_dataset.html*)

Sensor	Bandwidth (μ) Full Width–Half Maximum							
	Band 1	Band 2	Band 3	Band 4	Band 5	Band 6	Band 7	Band 8
TM	0.45–0.52	0.52–0.60	0.63–0.69	0.76–0.90	1.55–1.75	10.4–12.5	2.08–2.35	N/A
ETM+	0.45–0.52	0.53–0.61	0.63–0.69	0.78–0.90	1.55–1.75	10.4–12.5	2.09–2.35	.52–.90

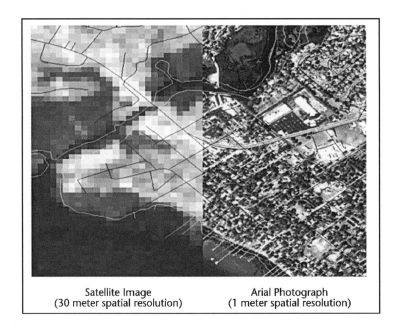

Satellite Image
(30 meter spatial resolution)

Arial Photograph
(1 meter spatial resolution)

FIGURE 9.6. Comparison of spatial resolutions for aerial photography and Landsat 7 and most Landsat 8 bands.

resolution, and two thermal infrared bands are interpolated to generate 30-meter resolution from data collected at 100-meter resolution. The resolution does not mean that an object of that size can be *consistently* detected and identified. Various atmospheric and situational characteristics play into this. You might think of this as simply the measure of the side of one of the raster cells detected by the remote sensing technology.

Temporal resolution depends greatly on the spatial resolution of the sensing technology. High spatial resolutions will record a great amount of data for a small area, requiring much longer to return to a place than low spatial resolution sensors. For example, Landsat 7 with 30-m spatial resolution revisits a place only once every 16 days. The Advanced Very High-Resolution Radiometer (AVHRR) has a spatial resolution of 1.1 km and revisits a place once every day.

Spectral resolution is an important characteristic. A coarse spectral resolution inhibits the ability to detect certain wavelengths and to distinguish features. A finer spectral resolution sensor can be used to represent different features based on the distinct wavelength patterns detected by the sensor.

Classification

Considering these three types of resolutions and other sensor and environmental characteristics, an operator can make a choice about how to classify the pixels from a scene using either supervised or unsupervised classification. **Supervised classification** means that the operator participates in an interactive process that assigns pixels to categories. **Unsupervised classification** occurs automatically without instructions from the operator.

Types of Sensors

This discussion of principles focuses on satellite-based remote sensing technology. This is only part of the available remote sensing technologies. The same technologies used for satellites, or adaptations thereof, are often used for remote sensing technologies used by airplanes, helicopters, and in some cases handheld formats.

Photography

Photography is the most common remote sensing technology. In fact, some of the first military remote sensing satellites used cameras with film in the 1960s. The film was dropped out of the satellite in a special heat-resistance reentry container with a parachute and while falling was picked up in the air by an airplane. Satellites still use cameras, but most of the images are now captured and stored digitally. Satellite sensor technologies using photography are "panchromatic" or sensitive to the full visible spectrum. The potential resolutions of photographic images are very high, but may be limited by data acquisition costs. Many governments and companies use aerial photography as a means of data collection. Using ground reference points and calculations to remove subtle changes in the airplane's movements, two aerial photographs taken simultaneously can be used to make a stereoscopic image. They are a very useful type of remote sensing because when viewed with some additional equipment like a stereoscope, it is possible for most people to distinguish heights and elevation changes. A single photographic image that also has the effects of elevation change removed (called planimetric) is called an **orthophoto** and is georeferenced to a coordinate system.

Infrared

Usually when we refer to photographic remote sensing we mean recording EMR in the visible wavelength spectrum, but this can be broadened to include infrared. This can be done with the chemical applied to photographic film (called an emulsion) or by using digital devices built and calibrated to detect this EMR spectrum.

Multispectrum

The data collected and images made with Landsat, SPOT, and similar sensing technologies are known as multispectrum because they include different bands. The variability of multispectrum remote sensors opens up a vast number of application possibilities.

Hyperspectral

This type of sensor technology collects narrow spectral bands over a continuous spectral range for all pixels of a scene. For example, the Hyperion sensor, installed on the EO-1 satellite, collects 220 bands from blue to shortwave infrared in equal steps (from 0.4 to 2.5 µm) with a 30-m spatial resolution. Flying in formation with Landsat 7, data from the Hyperion can be used with Landsat 7 images and data.

Radar

Radar is an important remote sensing sensor type. Its ability to penetrate through cloud cover and into the ground make it very useful for applications in areas with frequent cloud cover and for geological work.

Sonar

Sonar is used for remote sensing applications underwater. Well known from the pings in submarine movies, high accuracy versions produce detailed data for bathymetric mapping of sea and lake floors.

Laser (LiDAR)

Not used on satellites, but on planes, helicopters, and from the ground, LIght Detection And Ranging (LiDAR) uses a laser to generate light pulses, the same way radar uses radio waves. LiDAR is highly accurate and cost-effective for surveying and mapping, collecting elevation data and even bathymetry along with techniques to support many applications. Because of LiDAR's speed, small units have been introduced to quickly scan an area—for example, a crime or accident scene.

Applications

Images acquired by satellites have been used to produce local, regional, national, and global composite multispectral mosaics. They have been used in countless applications including monitoring timber losses in the U.S. Pacific Northwest, establishing urban growth, and measuring forest cover. Remote sensing images have also been used in military operations, to locate mineral deposits, to monitor strip mining, and to assess natural changes due to fires and insect infestations.

Data Collection in General

Thinking about remote sensing in a most general sense, we can easily distinguish types of data collection by the platform and by sensor technology. If the remote sensing is based on satellite images or data, in most cases we are likely to have multispectral, hyperspectral, or radar images or data. If it is airplane-based, then we are more likely to have aerial photography, multispectral, or LiDAR images or data. If it is ground-based, then we are most likely to find photography, multispectral, or LiDAR images and data. These rules of thumb have exceptions, of course, and will change as certain types of sensor technology and remote sensing systems become cheaper. They are simply helpful in seeing the relationship between costs, types of data, and application types. Applications in smaller areas tend to use airplane-based or ground-based sensor technologies; larger areas tend toward satellite-based remote sensing.

Coastal Monitoring

An important application area is coastal monitoring. Because of the key role of dynamic processes in coastal erosion, coastal monitoring applications tend to use remote sensing sources that can repeat their observations often. Aerial and LiDAR photography and data may be suitable for smaller areas if the area is generally cloud-free; multispectral satellite images and data may be useful for larger areas, and radar may be used for large areas, or areas with frequent cloud cover (see Figure 9.7).

FIGURE 9.7. Multispectral sensors produce data and imagery to help monitor and model complex coastal changes.

Global Change

With an increase in average temperatures worldwide, shrinking glaciers, and shrinking ice packs, the study of changes to glaciers and Arctic and Antarctic ice fields has benefited greatly from the use of remote sensing images and data. The frequency of observations helps scientists keep track of changes to ice fields and even icebergs in the water. Detailed observations, combined with measurements on the ground, help researchers monitor minute changes in ice fields. Made available online to other researchers, these measurements, images, and data have become a crucial part of a key area of global change research (see Figure 9.8).

Urban Dynamics

Because of the frequency of observation, satellite-based remote sensing images and data have proven to be very useful in documenting and assessing the growth of large cities around the world and distinguishing changes and processes. Urban dynamics are complex, but individual changes in a single area can be compared to assess the impacts of various policies and urban planning programs. Data and models developed to understand past growth can be used to make predictions of future growth and to assess alternative policy and planning proposals. Figure 9.9 shows development at an appropriate level of detail for such a purpose.

Precision Farming

Detailed remote sensing images and data, from a variety of platforms, are used by farmers to reduce the use of and become more efficient in the application of fertilizers and pesticides. Agricultural factors including plant health, plant cover, and soil moisture can be monitored with remote sensing

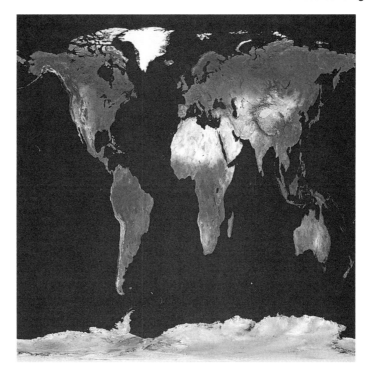

FIGURE 9.8. A composite of different multispectral data to produce a "picture-like" image of the world.

FIGURE 9.9. Aerial imagery (here from a digitized aerial photograph) can show a great amount of detail. Some newer satellites have been able to record data at similar resolutions, but have been restricted by national laws and security concerns.

data (see Plate 8). By combining the remote sensing images and data from different sources, deficiencies of one remote sensing system can be made up. For instance, Landsat provides multispectral data on average only once every 16 days for any place in the continental United States and is impaired by cloud coverage, even partially cloudy weather. By using radar data, scientists have been able to help farmers keep track of changing soil and plant conditions more frequently, which is especially critical during particular phases of plant growth (e.g., pollination).

Summary

Remote sensing is the collection of data without directly measuring the object. It relies on the reflectance of natural or emitted EMR. It has become an important and, in some applications, key means of data collection. The many types of remote sensors can be basically distinguished into two groups. Passive sensors rely on natural EMR; active sensors require an additional source of EMR. Remote sensing involves the complicated calibration of spectral signatures indicative of things or events with various characteristics and capabilities of sensors. Most sensors distinguish the EMR they detect as bands, which refers to specific ranges of EMR a sensor detects. Sensors also distinguish between spatial, temporal, and spectral resolutions. The number of applications using remote sensing keeps growing. Increasingly, with increasing accuracy and lower costs, remote sensing has been making inroads into traditional data collection applications.

REVIEW QUESTIONS

1. What does the term *LiDAR* stand for?

2. What does the term *panchromatic* mean?

3. Data from remote sensing is a powerful tool for analysis. What has prevented it from being more widely taken advantage of?

4. What is the longest operating earth observation satellite system that is still in use?

5. What is the size/scale of geographic area most well suited to being studied using remote sensing data? Small, medium, or large?

6. What remote sensing technology is being used in modern surveying?

7. What were some early applications for radar-based remote sensing?

8. What is the highest resolution of panchromatic remote sensing data now available?

9. When was remote sensing first used?

10. How is remote sensing data usually stored?

ANSWERS

1. What does the term *LiDAR* stand for?

   LiDAR stands for Light Detection and Ranging.

2. What does the term *panchromatic* mean?

   Panchromatic is a descriptive term for all wavelengths of the visible spectrum.

3. Data from remote sensing is a powerful tool for analysis. What has prevented it from being more widely taken advantage of?

   Remote sensing data has been costly to produce.

4. What is the longest operating earth observation satellite system that is still in use?

   The longest operating earth observation satellite system is called "Landsat." The first satellite of this system was launched in July 1972.

5. What is the size/scale of geographic area most well suited to being studied using remote sensing data? Small, medium, or large?

   Generally, remote sensing data is most useful for studying large areas.

6. What remote sensing technology is being used in modern surveying?

   Surveying now uses laser sensors for accurately measuring distances.

7. What were some early applications for radar-based remote sensing?

   Some of the first applications for radar-based remote sensing were climate analysis, iceberg detection, and geology.

8. What is the highest resolution panchromatic remote sensing now available?

   The highest generally available resolution is less than 1 m for panchromatic remote sensing data.

9. When was remote sensing first used?

   Most people consider the use of the telescope in the 17th century to be the first use of remote sensing.

10. How is remote sensing data usually stored?

    Remote sensing data is usually stored on computer using a raster format.

Chapter Readings

Conway, E. D. (1997). *An introduction to satellite image interpretation*. Baltimore, MD: Johns Hopkins University Press.

Gibson, P. J. (2000). *Introductory remote sensing: Principles and concepts*. New York: Routledge.

Lillesand, T. M., Kiefer, R. W., & Chipman, J. W. (2004). *Remote sensing and image interpretation* (5th ed.). New York: Wiley.

Sabins, F. F. (1997). *Remote sensing: Principles and interpretation* (3rd ed.). New York: Freeman.

Web Resources

⟳ This PBS Nova program offers a splendid introduction to remote sensing and is also online at *www.pbs.org/wgbh/nova/space/earth-from-space.html*.

⟳ NASA provides many fascinating images at its website at *http://visibleearth.nasa.gov*.

⟳ For information about Landsat 7, see the website *http://landsat.usgs.gov/index.php*.

⟳ Documentation of wetland destruction using animations is available online at *http://svs.gsfc.nasa.gov/vis/a000000/a002200/a002210/index.html*.

⟳ A detailed description of applying remote sensing in agriculture online is available at *www.nasa.gov/mission_pages/landsat/news/sweet-spot.html*.

⟳ A very insightful time lapse series of images to convey the process of Amazon deforestation is available at *https://www.youtube.com/watch?v=oBlA0lqfcN4*.

⟳ For in-depth discussion of everything related to remote sensing, with an emphasis on Landsat, but covering other sensor technologies in great detail, see *http://rst.gsfc.nasa.gov*.

⟳ For information about SPOT satellites, see *www.astrium-geo.com/na*.

⟳ For an excellent interactive tutorial about various aspects of remote sensing, see *http://satftp.soest.hawaii.edu/space/hawaii*.

⟳ An introduction to LiDAR is available online at *www.csc.noaa.gov/digitalcoast/_/pdf/lidar101.pdf*.

⟳ Download Landsat data online at *http://landsat.usgs.gov/LSDP.php*.

EXERCISES

1. Uses of Remote Sensing

Based on the discussion in the textbook and lecture presentation, determine with your neighbor three remote sensing applications and the data required for each. What is the spectral, the temporal, and the spatial resolution that each application requires?

2. Remote Sensing Laboratory Exercise

Objectives

To better understand the use and types of remote sensing.

Overview

For your first employment following completion of your undergraduate degree you can choose between helping a company set up a remote sensing service for urban areas wanting quick information about changes in their areas and helping its sister company provide the same service for national parks in the United States. Choose one of the positions and then make an assessment of the remote sensing data that is available for that service.

You are one of many people preparing a description of a new service. The management will review the descriptions and decide which services it will develop further.

Instructions

This is an important first step for either service and you have to be sure that your assessment is well documented. Your assignment is to prepare a one-page (single-spaced) assessment of the type of service you propose the company will provide using this set of topics as an outline.

1. Describe the service, including the remote sensing systems.
2. Explain the service in terms of the problem or issue you find requires this service.
3. Identify the data requirements (resolution and frequency).
4. Discuss the role of resolution (spectral, spatial, and temporal) in your study.
5. Identify any additional data the service will need.
6. Explain how this service can be developed into other services.

CHAPTER 10

Locations and Fields: Discrete and Nondiscrete GI

GI can be changed in many ways, whereas maps are usually very difficult to change—for example, copying a small map showing major cities of North America on a single sheet of office paper to a wall-size poster. Certainly, GI, like maps, is a form of geographic representation based on measurements, observations, and relationships (see Chapter 2). But unlike maps, GI has not been altered through a cartographic representation, although choices related to planned cartographic representations can influence how GI is stored. But, and its most important distinction vis-à-vis maps, GI can be transformed easily in various ways. John Sinton's distinctions between raster and vector data structures in terms of how they measure or control space (see Chapter 2) offer one way to understand key differences. This chapter takes a complementary approach considering key differences in location-based and field-based approaches to record GI. How the types of data structure and conventions constrain and influence these transformations in both overt and subtle ways is covered in Chapter 13.

We can start considering GI by differentiating three GI representation types: positions, networks, and fields (see Figure 10.1). The key issues for GI and cartographic representation related to the different GI types and transformations between the GI types are the focus of this chapter. Things and events can be represented as all of these types. Each offers different possibilities for recording measurements, observations, and relationships. When using a position, a thing or event is represented as a discrete record of location and properties, either in vector or raster format (see Figure 10.2). Not only is the thing or event fixed in space with a recording of its position in a coordinate system, its properties are also recorded based on observations and measurements made at or of that position. This type of GI representation remains the most common. GI networks represent things and

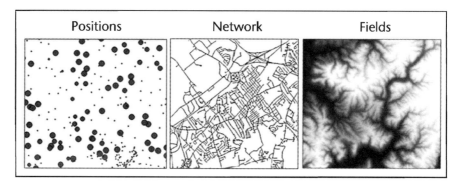

FIGURE 10.1. Examples of each type of GI.

events presented as an ordered arrangement of connecting points, called nodes (usually), lines, and sometimes areas. The things and events must be recorded in association with one of these geometrical network elements. Fields are used for nondiscrete things and events, which include anything that can be observed or sensed, but usually does not have clearly known or identifiable limits—for example, ozone, CO_2, or soil pH.

Each GI representation type is recorded using a database (see Chapter 7) or special computer-based storage created and maintained by software. This storage provides various ways to organize and index the GI on the computer. When using positions, topology is often used, also known as the georelational model, to control for errors and to make maintenance of the data much easier (see Figure 10.3). Positions usually best correspond to cartographic representation used for a map. Certain applications are more likely to use networks for transportation scheduling and routing or fields for environmental modeling. The storage of GI reflects the chosen GI representation types, but when not, it is possible to transform GI between various representation types.

Its ability to transform GI is the underlying reason why GIS has become such a worldwide success for so many human endeavors. It is always possible to transform one representation type to another using a GIS. For example,

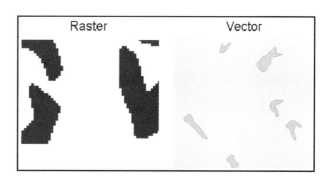

FIGURE 10.2. Examples of raster and vector GI representation types.

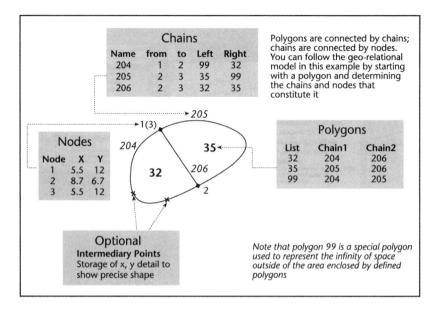

FIGURE 10.3. Key components of the georelational model, chains, nodes, and polygons.

cities shown as points in a position representation (Figure 10.1) can be transformed into areas in a field representation. The transformation into a field can also take into account that cities have fuzzy boundaries, not the sharp edges of a line.

The type of GI representation has consequences for quality and accuracy. Transforming the GI representation of cities as points into areas in a field may make it possible to show how cities diffuse into the surrounding area, but cities, even in a fuzzy form, are never perfectly round like the circle that represents them at small scales. Generally, we can distinguish between intra- and interunit qualities in relation to the geographic information type. Intraquality describes how well the differences between properties of units are represented—for example, population of cities in the categories less than and greater than 100,000 or with the exact count of the population. Interunit quality refers to the reliability that things and events are accurately represented—for example, the extent of a forest or a marsh. The boundaries of a wetland or city created at large scale will be much more accurate than a small-scale state map showing the location of wetlands and cities.

GI Representation Types

The three types of GI representation refer to concepts used by most GIS to represent things and events. Each representation type uses specific storage and indexing formats for recording the GI representation with information-processing technology. This section introduces each representation type,

IN DEPTH Topology

A modern branch of mathematics with great impacts in many fields, topology has been an important influence on the development of GI. Topology was introduced by one of history's greatest mathematicians, Leonard Euler, in 1736 when he published a paper on how to solve a puzzle that had perplexed residents of Königsberg (now Kaliningrad). The puzzle sought a solution about how to cross seven bridges that connected two islands in the middle of the city without crossing any bridge twice (see Figure).

Euler's solution was to abstract the problem into a set of relationships between vertices (also called nodes), edges, and faces. This is called a graph. Euler established that a graph has a path traversing each edge exactly once if exactly two vertices link an odd number of edges. Since this isn't the case in Königsberg there isn't a route that crosses each bridge once and only once.

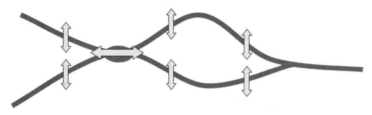

▨ Figure showing Euler's seven bridges of Königsberg problem.

The mathematics of this relationship are simple. To determine if there is a single relationship, count the number of vertices connecting three edges. If the number of vertices is two, then there is a single way around. Otherwise, at least one vertice must be crossed twice.

Euler contributed an immense body of work, over 775 papers, half of which were written after he went blind at the age of 59. The Königsberg problem is related to Euler's polyhedral formula, which is the basis for determining topology in a GIS:

$$v - f + e = 2$$

where v stands for the number of vertices, f for the number of faces, and e for the number of edges. Regardless of the type of polygon, this number will always be 2.

Topology was extended by numerous mathematicians in the late 19th century, and although most people learn little about it, it has been immensely significant for many technological developments.

Topology focuses on connectivity. In regards to GIS, topology is important for three reasons. First, it can be calculated to determine if all polygons are closed, lines connected by nodes, and nodes connected to lines. This allows for the determination of errors in digitized or scanned vector data. Second, it can be used in network GI to determine network routing. Finally, because it allows that the same line (edge) is used for neighboring polygons, the number of lines stored in a GIS can be greatly reduced.

discusses how it is used to represent things and events, and explains how, in very general terms, it is stored in a GIS. This section also describes topology, a foundation for vector GIS.

Position-Based Geographic Representation

Most GI is recorded using a position-based representation as points, lines, or areas (also known as polygons). This type of GI representation corresponds to the geometric primitives used to draw two-dimensional map elements. It is a handy and convenient way to create GI based on existing maps and for people used to working with maps. It is also very useful for many types of analysis (see Chapters 15–17). Of course, it can be transformed to other GI representation types.

Reflecting cartographic legacy, positional GI representations are usually two-dimensional and static. Events can only be shown in terms of positions and characteristics at a certain point in time. Measured properties are (1) either recorded as attributes of a spatial object, (2) are defined by the extent of the property, or (3) are associated with the measured properties of a predefined area (raster). Relationships either are defined by associations between attributes or are relationships that can be established and analyzed by transformations. The two most common storage techniques for this type of representation are vector and raster (see Chapter 2 for a more in-depth discussion).

Animation can be used to show events with position-based GI representations, but it is always based on a series of static geographic representations. Animations that show a series of images, just as frames in a comic, are relatively easy to create and show. However, they may be based on the interpolation of specific changes rather than measurements, which lessen their accuracy.

Vector GI is stored in a variety of ways. The most common format has been what people refer to as the **georelational model**. A special database, constructed and maintained by GIS software, uses very fast tables with topology. For many years the only "reliable" way of storing vector GI, this model is being replaced by proprietary database storage formats. Although the use of databases is expensive and usually requires specialized organization of the GI and work, databases are much quicker than the georelational model storage. However, because of its flexibility and robustness, the traditional georelational model should remain a commonplace fixture of GIS for some time.

The georelational model is extremely useful because it relies on topology. Topology provides a means to speed up many processes, information to check for polygon and line errors, and a way to reduce the storage requirements for GI.

The georelational model consists of three main components connected topologically. All three components are present and are linked to each other. The first component is a table with a list of polygons (or areas). It records the internal number of a polygon and the chains in the order that make up the polygon's boundary. The second component is the table with a list of chains (also called "lines" or "arcs"). Each chain entry consists of information about the polygons to either side of the chain and the start and end nodes of the chain. The start and end nodes define the direction of the chain and which polygons are left and right. The third component of the georelational model is a table of nodes. This table consists of the node identifier and the x and y coordinates of each node (see Figure 10.3).

Additional data beyond the three components of the georelational model can be made to improve its geographic representation and the cartographic representation, especially the addition of additional points used to define the precise shape of a chain and indexes to speed up queries and the drawing time.

Raster GI representation divides the data into a regular grid of equal size cells in commonly used GIS software. It can correspond to remote sensing raster data or field-based data, but the boundaries between cells of position-based GI reflect actual or interpolated boundary locations. A rasterized topographic map is a good example. Raster GI relies on various types of encoding to reduce the amount of storage required by a computer. If each raster or pixel cell is stored individually, the files become very large. A simple way to reduce the required storage (and one of the oldest) is to process each row of the raster data set from left to right, recording only when the attribute value changes and the number of cells following the change to the right. For example, if a row is 100 cells long and cells 1–20 have the value 156, cells 21–78 have the attribute value 123, and cells 79–100 have the attribute value 156 again, the run-length encoded (RLE) raster storage would only store 156:20; 123:59; and 156:21. Other systems are more complicated, but even more efficient. One of the most interesting storage formats is the quad-tree format which works like the RLE approach, but puts areas into a hierarchy based on cell values (see Figure 10.4). For example, an agricultural raster data set representing types of crops could distinguish crops at the highest level by the genus, at the next level down in the quad-tree hierarchy it could show the Linnean classification family, and at the third level of the quad-tree it could show individual species. The quad-tree is very efficient and very fast, but changes to the hierarchy can be very complicated and require a great amount of processing.

Network-Based Geographic Representation

The network geographic representation type is usually considered to be a subtype of the position-based GI type, but is distinct because of its special properties for representing topological relationships (see Figure 10.5). The commonality is that locations of things and events in networks are usually recorded using positions from a coordinate reference system.

The network geographic representation type uses nodes and links to indicate locations and connections, usually transportation or utility networks. These correspond to nodes and chains in the vector position-based geographic representation type but their function is different. The network can also store information about possible connections (e.g., possible turns at

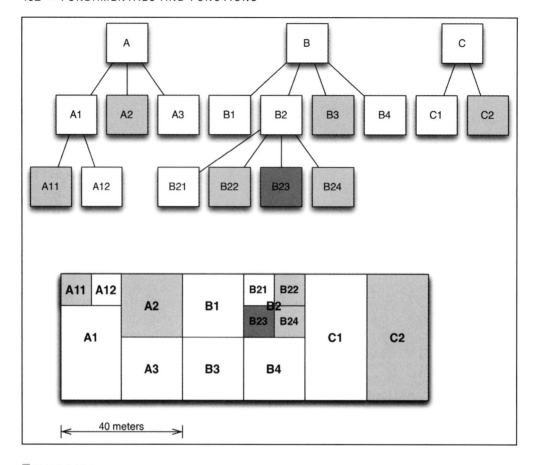

FIGURE 10.4. Simplified illustration of a quad-tree. The top part shows the optimized hierarchical structure for raster data with a 5 m resolution. The bottom part shows the data as a graphic. The quad-tree optimizes storage by maximizing the extents of collated raster cells sharing the same attribute.

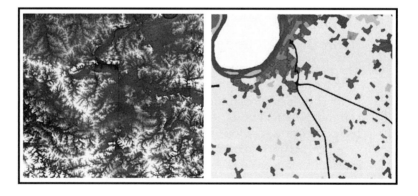

FIGURE 10.5. Two examples of field GI. On the left a DEM, on the right GIRAS land use.

an intersection), and links can store information about how nodes are topologically connected (e.g., Chicago is connected to St. Louis by Interstate 55). Topological information is essential for network GI.

Nodes can be added with coordinates from a coordinate system and with additional points with coordinates to define the shape of the networks, for example, situating Chicago and St. Louis on the map in a geographically correct arrangement. However, many networks will be cartographically represented without location information, allowing the map to be very simple and easily read (e.g., public transportation maps). (See Plate 9, the London Underground Map.)

Field-Based Geographic Representation

For the representation of nondiscrete, mainly environmental, properties including soil moisture, soil pH, or the distribution of airborne particles and substances, for example, ozone, dust, or pollen, fields are the ideal GI representation type (see Figure 10.5).

Conceptually, fields are nondiscrete, meaning no precise and accurate boundaries can be made between soil pH 6.7 and 6.8, and the properties of a field can be modeled using geostatistical techniques that take geographical relationships into account. In other words, a field can be represented as an *n*-dimensional expression. For practical reasons, the storage of the field GI representation type most frequently uses raster data structures. The nondiscrete character of the field should always be considered when working with these rasters. It is easy, but wrong, to interpret raster cell boundaries of field data as the sharp boundaries between different attribute values, when, in fact, the geographic things and events represented by a field are continuous properties.

A triangular irregular network (TIN) is a specific format for the representation of fields that relies on a network of lines connecting sampled points with known values (see Figure 10.6). The connections form a Delauney triangulation, which means that each point is connected to only two other points to create triangular faces. This type of GI representation is most commonly used for the visualization of elevation data, but can be used for any data that is collected using irregular samples in an area. Dynamic versions of TIN make it possible to rapidly change the TIN. The changes can be so rapid that dynamic TIN holds the potential to help train people for complex navigation situations (see Figure 10.6).

Transformations

Transformations are key to the abilities of GIS, and this is especially true for working with field-type GIS. This type of geographic representation may not originate with data collected for every point in the area of the field. Since this detailed data collection would be practically impossible, alone due to costs, most field data is usually the result of transforming a smaller, but hopefully well selected position-based group of GI observations and measurements. For instance, a property of soil, pH, shown as a continuous field type of GI

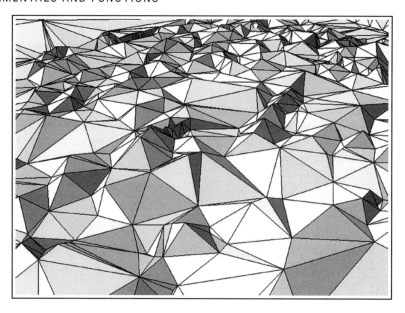

FIGURE 10.6. Example of a TIN data structure. Each triangle is a facet of a hill slope representing a change in elevation, orientation, or the relationship between these two characteristics.

may be based on an interpolation of soil samples collected at various points. The soil pH field type GI can be transformed back into a position-based GI representation as contours to indicate where soil pH changes (e.g., a contour for every 0.5 change in soil pH). Transformations can be applied to any representation of GI. GI can be transformed to different types—for example, positions to fields, or networks to positions, or from one position-based GI representation to another (e.g., points to lines).

The transformation concept goes back to Waldo Tobler's development and application of the mathematical transformation concept to cartography. For Tobler, the map is more than a cartographic representation of GI; it is a device for storing information. Tobler worked on mathematical techniques and analytical methods to transform maps into forms of information that can be changed further. Thanks to Tobler's conceptual work, we regard GI not just as data, but as data with meaning, which can be transformed and combined with other GI to create new forms of GI. With the transformation concept comes an understanding of GI as sets of associations with particular representations that can be converted to create other sets of associations.

What Are Transformations?

Transformations are operations on GI that change the information content by manipulating GI and changing it into other GI representation types. Many transformations are frequently used GIS operations. For example, a buffer operation can transform a point that represents a well into a polygon that represents the zone around the well unsuited for locating an underground fuel oil tank. This zone can be represented as positional or field GI, depending on the operation chosen. The zone can be transformed into the other GI representation types. Transformations of GI can also change attributes. An example of an attribute change is converting temperature recorded in degrees Celsius to degrees Fahrenheit. In both cases, the key change involves transforming the GI representation. What information is measured for a point, such as a well, is only of limited validity for an area, such as a theoretical plume extent. A transformation can produce new GI based on calculations that show a relation, as in the example of a buffer.

Examples

The two most fundamental GIS operations, buffers and overlays, are examples of GI representation transformations. Buffers transform position-based GI into other types of position-based GI or fields. Overlays transform two position-based GI data sets into one. What these operations involve and how they transform GI demonstrates the key role of transformations for GI and its much greater usefulness compared to maps.

Buffer Transformations

A buffer transformation is the simplest transformation to grasp, but its operation can actually be quite complex. Practically, based on the position of one or more GI objects, it determines the zone around the objects using one or more distances. Figure 10.7 (left) shows a simple 100-foot buffer around a well. But what do the 100 feet (about 30 m) represent? They may simply be the regulatory zone where no animal waste disposal is allowed. But that regulation could be based on more complex geographical relationships. Maybe the 100 feet corresponds to the well recharge zone calculated using a hydrological model that considers both the soil type and geology common in the

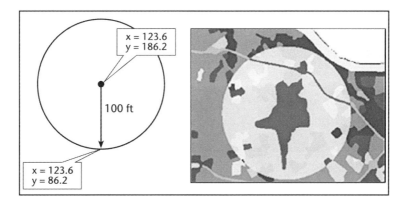

FIGURE 10.7. A 100-foot buffer around a point representing a well produces a vector area or field (left); buffered zone of land use around the Cincinnati/Northern Kentucky Airport (right).

area. The areas of buffers usually are used to show a geographical relationship. Based on an understanding of the relationship, distances are used to show the extent of the relationship. This technique is used to indicate area affected by vehicle or airplane traffic. Complex models may only use buffers to represent the results of calculations that work with fields and model things and events in terms of relationship vectors. This simple operation is a very powerful transformation. In all cases, obviously the accuracy and quality of a buffer depend on the underlying model and explicit (or implicit) assumptions.

Overlay Transformations

GIS overlay is, depending on who you speak with, the first or second most important operation for GIS. Either way, it is without doubt one of the most significant operations. It is also one of the primary transformations, but the transformations performed by an overlay depend on the type of GI representation (see Figure 10.8).

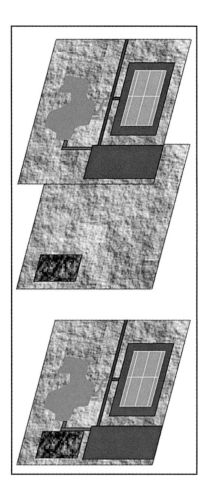

FIGURE 10.8. Overlay transforms by combining two (or more) data sets based on the location of features in a coordinate system.

Positional GI combines the geometries (points, lines, or areas) of two data sets based on a common coordinate system. The geometrical transformation is only the geometric process of determining the intersections between objects from each data set and the assembly of new objects that correspond to the original objects. Attributes from the original objects are assigned to the new objects based on the location of the original objects. The attribute transformation begins at this point. Various operations, logical and mathematical, are used to transform attributes and relate them—for example, evaluating soil type and soil moisture to determine crop suitability. Raster GI performs these attribute transformations. This assumes that both raster data sets use the same raster size and origin point (otherwise some complex geometric transformations must first take place). Chapter 16 covers these issues and the overlay operation in more detail.

Summary

This chapter examined GI representation types and transformations. GI representation types are the formats available for GI: positions, networks, and fields. Positions and networks rely on vector data formats; fields rely on raster data formats. Positional GI is stored in a GIS as points, lines, or areas (also known as polygons), most often following the georelational model that uses topology. Networks also use these data formats, but areas are of very limited use in a network. Points, called nodes in networks, are much more important.

Transformations are operations on GI representation types that change the information content. A buffer transforms a point through a distance measure into an impacted area.

REVIEW QUESTIONS

1. What sets GI apart from maps in terms of discrete and nondiscrete information?

2. Why are multiple types of data structures needed?

3. What is Tobler's transformational concept?

4. What is the main difference between discrete and nondiscrete GI?

5. What is the main difference between topological and nontopological vector data?

6. What is a quad-tree?

7. What is a triangular irregular network (TIN)?

8. How can the GI storage format impact GI representation?

9. How does a buffer operation transform a geographic representation?

10. Why can't maps be transformed?

ANSWERS

1. What sets GI apart from maps in terms of discrete and nondiscrete information?

   GI offers multiple ways to store and transform data that can be used to make meaningful representations of things and events as GI. Maps can show both discrete and nondiscrete information, but the information cannot be transformed.

2. Why are multiple types of data structures needed?

   Different types of data structures make it possible to adequately geographically and cartographically represent observations of things and events.

3. What is Tobler's transformational concept?

   Tobler's transformational concept is the development and application of the mathematical transformation concept to cartography. With this concept comes an understanding of GI as sets of associations with particular representations that can be converted to create other sets of associations.

4. What is the main difference between discrete and nondiscrete GI?

   Discrete GI shows things with fixed boundaries; nondiscrete GI shows processes or states of processes.

5. What is the main difference between topological and nontopological vector data?

   Topological vector data has a set of relationships between nodes and links; nontopological vector data maintains only start, possibly intermediate, and end points.

6. What is a quad-tree?

   A quad-tree is a data structure for the efficient storing of raster data following a hierarchy based on areas of contiguous attribute values.

7. What is a triangular irregular network (TIN)?

   A TIN is a data structure for storing GI based on distance relationships and single values; it is most widely used for storing and modeling elevation data.

8. How can the geographic data structure impact GI representation?

   It allows certain attributes and relationships to be better stored than others; transformations make it possible to convert GI to other formats that may resolve the limitations with one particular type of data structure.

9. How does a buffer operation transform a geographic representation?

Based on existing geometry (point, line, area) and attribute value(s), it creates a new area that represents a new geographic representation with a new thing or event.

10. Why can't maps be transformed?

Maps cannot be transformed because of the cartographic representation and recording in the fixed media of a map. Maps cannot be directly transformed into other representations. Information collected from maps through digitization can, however, be transformed.

Chapter Readings

The second edition of this text contains a wealth of new and additional information, but the first edition is still a classic. See:
Burrough, P. A. (1987). *Principles of geographical information systems for land resource assessment.* Oxford, UK: Oxford University Press.

This book presents the use of databases for representing GI:
Rigaux, P., Scholl, P., et al. (2002). *Introduction to spatial databases: Applications to GIS.* San Francisco: Morgan Kaufmann.

From the computer science perspective, see this key book documenting the development of GIS:
Worboys, M. F. (1995). *GIS: A computing perspective.* London: Taylor & Francis.

Web Resources

↻ For an introduction to some of the fundamental GI representation issues, see the Wikipedia entry online at *http://en.wikipedia.org/wiki/Geographic_information_system*.

↻ For a paper that discusses some of the limitations of the widely used types of GI representation, see the website *www.ucgis.org/priorities/research/research_white/1998%20Papers/extensions.html*.

↻ A good GIS tutorial, from the South African Eastern Cape Province, can be found online at *http://linfiniti.com/dla*.

EXERCISE

1. Euler's Seven Bridges Problem

Description

Topology is a field of mathematics where distance is not relevant. In this exercise, you will examine some of the basic concepts of topology.

Exercise Instructions

On this rough map illustrating the seven bridges of Königsberg problem that motivated the mathematician Leonard Euler to develop topology, try to draw with your pencil in one continuous line a way to walk around the city crossing each bridge only once.

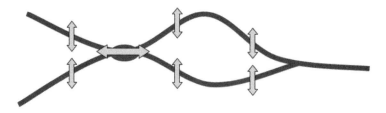

Questions

1. Why do you think this is so difficult?
2. Is it possible?

As a second case: assume a flood washes out one of the bridges in Königsberg, leaving six. Draw a route around the city now using one continuous pencil line.

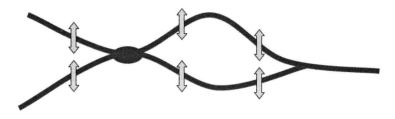

Questions

1. Does it matter which bridge you take away?
2. What if you add bridges?

EXTENDED EXERCISE

2. Networks, Topologies, and Route

Overview

In this exercise, you will use GPS equipment to determine the locations of several key points near campus, determine the time to walk between locations, and prepare a network graph where distance corresponds to time.

Objectives

Learn how to apply topography in geography.

Exercise Steps and Questions

1. Configure the Equipment

Make sure the GPS is working properly. Please check the battery status on the main menu (you get there by pressing the main menu button twice). If the power indicator is significantly below 25%, please see the instructor to get new batteries.

2. Collect Location and Travel-Time Data

In this step, you will need to collect locational data for each point and the time it takes to walk between each location. *You should plan on taking 1 hour to collect this data.* On the table on the next page, first write down the names of seven *places* (the first place should be in front of the main entrance to the building where class normally meets) you will collect location data for. Go outside the main entrance and wait until you have excellent GPS satellite reception (your accuracy should be less than 30 ft). Write down the coordinates displayed on the GPS receiver, the departure time, and then start to walk at a comfortable pace to your second location. When you get there write down the arrival time and location information. When finished writing this information, write down the departure time and proceed to the next point.

Please note:

- Each point should be at least 200 m from any other point—farther is even better.
- Each connection between points should only be recorded once.
- If you walk in the order of your locations, your arrival and departure times are always related. However, if you change the order, you will need to make a note of that on the worksheet.

	Description	Easting	Northing	Elevation	Arrival Time	Departure Time	Difference
1							
2							
3							
4							
5							
6							
7							

Comments/Observations:

3. Make a Map of the Locations

On a separate sheet of paper, prepare a drawing showing the geographic locations of the data you observed above and the routes you traveled to each point. Make sure to indicate scale and include a legend that explains the symbols you used.

4. Make a Network Graph of Locations and Travel Times

In this step, you will create a schematic drawing of the seven locations from Step 2 on a new sheet of paper. You should arrange the locations in a fashion similar to the graphs showing topology.

Because this network shows travel time, you want to show the distance between locations as a scale equivalent. Each location should be labeled. Make sure to determine the appropriate scale before drawing the graph: assuming the shortest travel time between locations was 5 minutes and the average travel time was 12 minutes, you want to have a scale that fits all your points on a 8.5″ × 11″ sheet of paper. If the maximum travel time is 30 minutes, a scale of 1″ = 5 minutes will need 6″ on the paper. *You should do this in pencil at first in case you need to make changes.*

5. Draw a Network without Scale

Based on the network graph from Step 4, draw a network graph that is not scaled, but only shows the connectivity between locations. You should still arrange the locations in a fashion similar to the graphs presented in the lecture on topology, but don't scale the distances by time.

6. Evaluate Your Network Graph

Using Euler's Characteristic,

$$v - e + f = 2$$

evaluate your graph from Step 5, where v is the number of vertices of the polyhedron, e is the number of edges, and f is the number of faces (remember that vertices are your locations and intersections of edges, edges are the connections between vertices, and faces are the areas bounded by connections). Note: Always add one face for the surrounding area of the network.

The value should be 2. If it is not, check your graph to make sure you have included all locations and added vertices where paths meet.

Questions

1. Is your graph all on the same elevation? If you need to consider multiple elevations—for example, to show overpasses or bridges—how would that change the connectivity of your graph? What is the term used for graphs that are on the same elevation or level?

2. Copy your scaleless network graph to another sheet of paper and indicate how you would traverse the network in a single trip. If you can't, indicate which node you would have to cross twice.

3. How long will it take you to get around your network once?

4. You have a geographic map, a scaled network graph, and a scaleless network graph for a portion of the campus. What is the map better for and what is the graph better for? What does the use of a scale add or detract? Is it necessary to show network connectivity? Please identify two activities for each map and each graph.

PART 3

Techniques and Practices

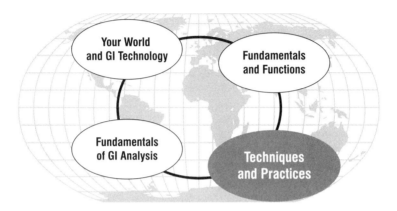

The chapters in this third part of the book, Techniques and Practices, move from the basic components and sources of data that comprise GIS (which were the subject of Part 2) to consider important topics of getting GIS up and running: how well will GIS perform their basic tasks, which are representing and communicating? Taking the raw materials that comprise a GIS, and shaping them, requires making decisions about simplification of data and symbol choice on maps. These are cartographic principles, as covered in Chapter 11, and since a GIS may contain, for example, epidemiological data overlaid on geographic, the chapter also covers different kinds of graphical displays of data that might "fly out" for users interested in epidemiological data, or a topographic view of a political map. Chapters 12 and 13 cover the way cultural forces influence GIS development and how GIS are used in public policy, the "administration" of spaces. Chapter 14 is new to this edition and addresses the centrality of location data in mobile devices and web and Internet mapping. These chapters can be read in sequence or portions selected to fit the scope of a course or workshop.

Part 3 lays the groundwork for Part 4 in which we use GI to analyze events happening in the world. In a very real sense, GI helps us make sense of the world.

CHAPTER 11

Cartographic Representation

Visualization and the general approach to the design of maps are critical to successful geographic communication, which has two components: (1) geographic representation considers how to abstract things and events from observations and measurements of the world (see Part 1); and (2) cartographic representation deals with the creation of visual techniques and forms following cartographic principles. This chapter covers the latter: principles of cartographic representation including scale, generalization, classifications, media formats, and cartographic presentation types. We go into greater detail than in the overview in Chapter 3. *Communication,* broadly understood, is how we convey or share representations of information with others. Depending on the means and modes of communication, different cartographic representations could be required. A classroom wall map of Europe or North America will show things and events differently than small maps of those areas presented on a TV news broadcast.

A cartographic representation is an abstraction from the geographic representation. Therefore, a key concern of cartographic representation is to perform that abstraction reliably. Good communication may require keeping the geographic representation in mind while developing the cartographic. Ultimately, the cartographic representation should maintain the integrity of the GI. Otherwise, substantial errors and distortions may occur. This chapter suggests that applying the principles of cartographic design pragmatically will result in a successful cartographic representation.

Maps and Visualizations

The focus of this chapter requires a distinction between maps and visualizations. Maps are well known to all: they are two-dimensional, or limited three-dimensional, static, and usually printed on paper or some material with many of the same properties of paper. **Visualization** refers to a broader concept, which is mainly connected to images, especially dynamic visualizations, made with computers. Connecting the concept of visualization to the type of media used in communication helps in distinguishing visualizations from maps at a pragmatic level, but a broader consideration would holistically consider both as part of a system that also relies on other technologies for visualizing geographic features and events. Two factors in visualization are: (1) **Display scale**, which is involved with the capabilities of electronic displays ranging from cell phone displays to various kinds of projectors. The size and resolution of these displays is often limited and may necessitate user-controlled zooming functions to change the focus of the display. (2) The tangibility and temporality of visualizations are often fleeting. What is on the display one second may be gone the next, and returning to previous images may be impossible or cumbersome.

Alan MacEachren, a cartographer, developed a conceptual framework for thinking through cartographic representation issues and the important roles of cartographic representation (see Figure 11.1). His work focuses on various forms of map use for cartographic communication or visualization as communication. All map use involves visualization, or cartographic representation, but map use also varies in terms of its relation to public and private spheres, human interaction with maps, and how the map is used for presentation or discovery. These aspects of cartographic representation should not only contextualize geographic representation, but should also be included among the factors considered when preparing maps or visualizations.

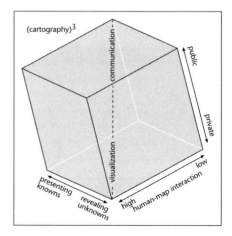

FIGURE 11.1. MacEachren's Visualization Cube.

IN DEPTH **Suggested Map Elements**

Cartographers would likely argue for days and nights about which elements are necessary in a "good" map. For pragmatic purposes, there are six elements that people agree are helpful, even if not always necessary:

- Scale indicators show the relationship between distance measures on the map and the actual ground distance. Scale bars are most common and practical. The representative fraction provides important information for experienced map users.
- Legend explains what the symbols used for a map mean.
- Title provides a simple description of the map, possibly also indicating potential uses and audience of the map.
- Author offers readers the name of the institution, group, or individual responsible for creating the map. It can help point to the acceptance of the map.
- Orientation (north arrow) helps people match direction to top, bottom, left, and right of the map they are using.
- Date indicates when the map was produced and may also suggest when the data was collected.

Explanations and contact information can additionally be included, if they are important for communication, or necessary to ensure that readers can find relevant information for more specific questions.

Note that these six map elements do not necessarily refer to visualizations. Although they can be considered for a visualization, the constraints on a visualization of size and length of display means that map elements should only be used when absolutely essential to communication.

Pragmatic Cartographic Representation Issues

For the pragmatic creation of GI and maps, you need to understand some basic issues and their roles for maps and visualizations.

Scale

The geographic area shown on a map must be at a scale. The selection of scale and the use of scale have many consequences (see Figure 11.2). The most important thing to remember is that large areas are shown with GI or in maps and visualizations at a small scale, while small areas are shown at a large scale. This may be at first glance counterintuitive; however, scale always states the relationship between one unit of distance on the map and the same unit of distance on the ground. Large scale refers to a large ratio between map and ground units; small scale refers to a small ratio.

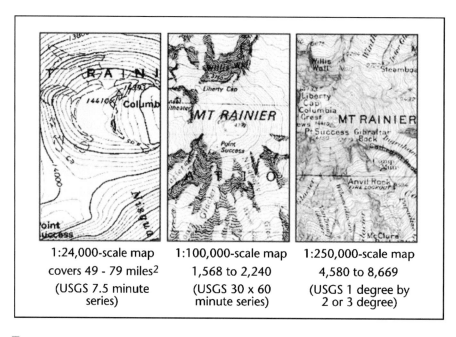

1:24,000-scale map	1:100,000-scale map	1:250,000-scale map
covers 49 - 79 miles^2	1,568 to 2,240	4,580 to 8,669
(USGS 7.5 minute series)	(USGS 30 x 60 minute series)	(USGS 1 degree by 2 or 3 degree)

FIGURE 11.2. Some map scale examples and sizes covered by USGS map sheets.

Scale for maps and GI, in other words, is usually expressed as a mathematical relationship. However, this is only one of three ways to express scale. Scale can also be expressed in words—for example, "1 inch equals 1 mile" or by using a scale bar (see Figure 11.3).

The representative fraction is the most important way to represent scale. It offers a clear indicator of the relationship between distance measurements on the map and distances on the ground and vice versa. For instance, many topographic maps from around the world are published at a scale of 1:25,000. The maps at this scale are almost always metric, which means that distance on the map and on the ground is measured using meters or a unit of measurement related to the meter. A word expression of scale of these maps is

Representative Fraction	One Inch equals	One centimeter equals
1:24,000	2,000 feet (exact)	240 meters
1:50,000	4,166 feet	500 meters
1:63,360	1 mile (exact)	633.6 meters
1:500,000	8 miles	5 kilometers
1:1,000,000	16 miles	10 kilometers

Scale Bar Example

1000   0   1000   2000 Miles

FIGURE 11.3. Some scale representations.

"one centimeter on the map equals 250 meters on the ground." Another word expression for this scale is "four centimeters equals one kilometer." In fact, for ease of reference, these maps were often called "four centimeter" maps. But the words are still long. Every time the map would be described, people would need to remember the phrase. If people weren't familiar with the phrase, a great deal of explanation would be needed. If we express the scale as 1:25,000, it's much easier. This means that 1 mm on a map is equal to 25,000 mm on the ground. There are 10 mm in a centimeter, so 1 mm on a map is also equal to 2,500 cm. With 100 cm in a meter, we can finally convert the relation of map measurements to ground measurements, or 1 mm equals 25 m. Knowing this relationship, we can easily calculate how long a bridge 4 mm on the map is on the ground. Knowing how many millimeters are in a centimeter, we can also easily calculate how wide a 6-cm field is—or any other distance for that matter (see Figure 11.4).

When using inches, feet, yards, and miles, the mathematics are more complicated, but the principles are the same. The main numerical relationship to know is that the 5,280 feet, or 1,760 yards, of a single mile are equal to 63,360 inches. In other words, at a scale of 1:63,360, 1 inch on the map is equal to 1 mile on the ground. The calculation of distance measurements from a map or visualization to the equivalent ground distance when using standard or imperial distance units must always first determine the number of feet on the ground in 1-inch distance measured on the map. For example, 1 inch measured on a map in the United States at the scale of 1:24,000 is equal to 24,000 inches on the ground, or 2,000 feet. If the distance between two intersections on a 1:24,000 scale map is 6 inches, the distance between them on the ground is 12,000 feet or a little more than 2.25 miles. You should note that this imprecision in reporting distances is just one of the many reasons for using the metric system for distance measurements. The ease of calculations is perhaps the second reason.

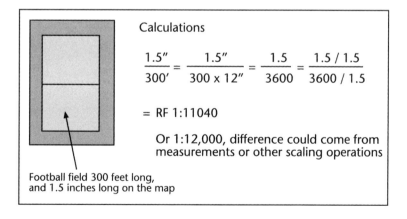

Scale can also be expressed by using a bar that indicates a single distance or multiple distances on the ground. A visual comparison offers easy approximation of distances, and the use of a ruler to determine the equivalent distances on the map makes it possible to determine the representative fraction and to do similar calculations.

Scale is also a very significant indicator of the detail shown by a map or visualization. Generally, smaller scales show less detail, while larger scales show more. Many cartographers would agree that maps at scales greater than 1:50,000 scale should be called small-scale maps, while maps at scales between 1:100 and 1:50,000 should be called large-scale maps. Although admittedly there is some disagreement among cartographers about this system, this division provides a meaningful starting point for the discussion.

Finally, different disciplines work with a distinct scale or possibly a set of scales. These scales are usually related to established disciplinary conventions. Because of the cost of GI and the standardization of topographic maps in most European countries, many European planning activities rely on scales of maps created by their national mapping agencies. For smaller areas, they rely on scales standardized for cadastral mapping. This is starting to change as GI becomes more common, but the 1:25,000 and 1:50,000 scales are still prevalent.

Determining Scale by Other Means

If you don't know the scale of a map or visualization, you can figure it out by using one of three techniques.

- Use of known features
- Use of lines of latitude or longitude
- Use of map object comparison

The use of known features uses the ground distances that are clearly known for an object on the map or visualization. Based on the measurement of the distance on the map for the same feature, the map scale can be easily calculated (see Figure 11.5). Because of distortions of paper or minute measurement error, often the results may be slightly in error.

The known distances between lines of latitude or longitude can also be used to determine scale (see Figure 11.6). This approach is especially suitable for small-scale maps. The distances between lines of longitude or latitude can be determined from tables published on the Internet.

Comparing distances between the same features on two maps when the scale of one map is known is the third technique. Needless to say, great care has to be taken when making measurements.

Generalization

To cartographically represent things and events, they must be abstracted to fit the target scale or scales and still accurately retain key properties and

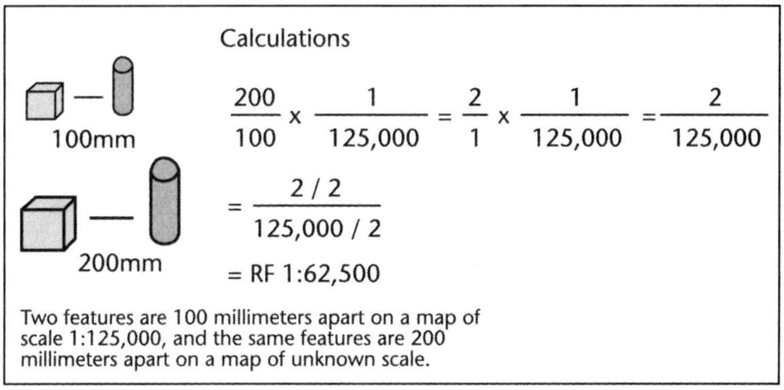

Calculations

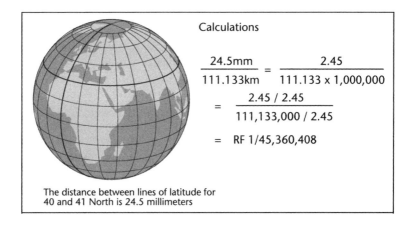

$$\frac{200}{100} \times \frac{1}{125,000} = \frac{2}{1} \times \frac{1}{125,000} = \frac{2}{125,000}$$

$$= \frac{2/2}{125,000/2}$$

$$= \text{RF } 1:62,500$$

Two features are 100 millimeters apart on a map of scale 1:125,000, and the same features are 200 millimeters apart on a map of unknown scale.

FIGURE 11.5. Map scale can be determined by comparing the distance between two features on two maps, when the scale is known for one map.

relationships (see Figure 11.7). Just as scale is an indicator about the amount of detail in a map, it is also an indicator about the amount of **generalization**, the abstraction of features to reduce complexity in maps or GI. Generalization serves to assure that cartographic communication works as well as possible at a particular scale or scales. First, scale defines the relationship between an area on the ground and an area on maps or in visualizations; second, scale constrains the possibilities for showing things, events, and relationships.

Generalization is often the most important part of cartographic representation because it assures the role of the map or visualization for the process of cartographic communication. It involves the alteration of GI and symbols. The underlying graphical issue is that with decreasing map scale, graphical elements (points, lines, areas) will refer to larger things and events.

Calculations

$$\frac{24.5\text{mm}}{111.133\text{km}} = \frac{2.45}{111.133 \times 1,000,000}$$

$$= \frac{2.45/2.45}{111,133,000/2.45}$$

$$= \text{RF } 1/45,360,408$$

The distance between lines of latitude for 40 and 41 North is 24.5 millimeters

FIGURE 11.6. Determining map scale using the distance between lines of latitude or longitude.

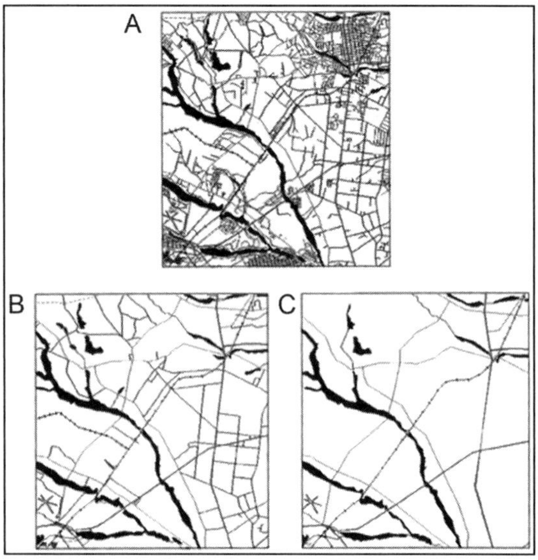

FIGURE 11.7. Generalization choices are important because they ensure that a map can be correctly used at the intended scale and for the intended purposes. Presented at the same size here, the scale of (A) is 1:100,000; (B) is 1:250,000; (C) is 1:500,000.

A square 1 mm on each side can represent a playground 25 by 25 m in size on the ground at a scale of 1:25,000. At a scale of 1:100,000, the same 1 mm square shows a playground 100 by 100 m large.

Two other issues make generalization necessary. Because the clear cartographic representation of things and events can take up disproportionate space at smaller map scales, generalization is necessary to avoid cluttering the map or visualization and to clarify important relationships. Generalization is also needed because of emphasis on certain things, events, or relationships that are central to the communication role of the map or visualization.

Work on generalization in cartography recognizes a number of different operations. The typologies of generalization operations vary, but certainly distinguish between points, lines, and areas. Other issues reflect institutional, functional, or conventional understanding of the cartographic representation and geographic representation that the generalization should support. To introduce the generalization operations, the five most commonplace operations offer sufficient insight into the significant role of generalization for cartographic representation and communication.

Aggregation

Aggregation is used to reduce complexity. It can either be used to merge features that border one another (e.g., land use types) or to group features that are nearby (e.g., buildings belonging to one factory). A potential consequence of aggregation is that detailed characteristics can be lost, important relationships homogenized away, or misleading presentations made (see Figure 11.8).

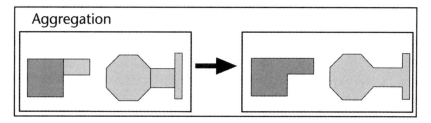

FIGURE 11.8. Simplified illustration of aggregation operation.

Displacement

Displacement involves changing the position of points, lines, or areas either to clarify important characteristics, to avoid conflict with other features, or to resolve clutter. It is commonly used for emphasizing the relationship of roads to other features on small-scale transportation maps. Because it results in changes to the positions of things and events, GI collected from small-scale maps needs to be carefully verified against larger-scale GI for the same area. The potential distortion of actual geographical relationships should be taken into consideration when working with any map or visualization (see Figure 11.9).

Enhancement

Enhancement exaggerates geometrical characteristics of lines or areas to clarify the geographic characteristics or help in inferring geographical relationships from the cartographic presentation. As with the displacement operation, it creates positional changes that distort the relationship between the features and actual things and events on the ground (see Figure 11.10).

Selection

Selection is used to retain certain points, lines, or areas for the cartographic representation and to remove others. This mainly aids in dealing with clutter,

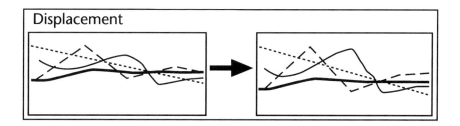

FIGURE 11.9. Simplified illustration of displacement operation.

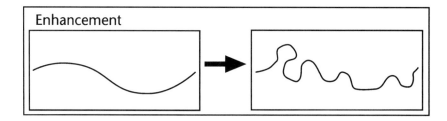

FIGURE 11.10. Simplified illustration of enhancement operation.

although this operation is also used to help clarify relationships by removing details. Naturally, this operation can result in significant distortions and possibilities to misinterpret geographic relationships (see Figure 11.11).

Simplification

Simplification reduces the amount of geometric detail for individual lines or areas. This can be done by arbitrarily removing points that describe the shape of a line or areas, or can be done with several algorithms, or can be done by hand. In each case, varying degrees of changes to the positions result, leading to possible distortion (see Figure 11.12).

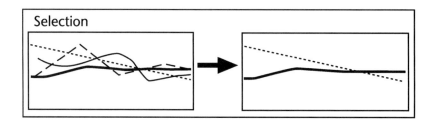

FIGURE 11.11. Simplified illustration of selection operation.

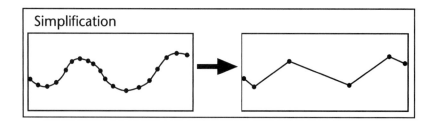

FIGURE 11.12. Simplified illustration of simplification operation.

Classification

The characteristics of things and events stored as GI, or associated with the corresponding GI, need to be clearly organized for communication. Humans can only understand a certain amount of information at once (Figure 11.13). The rule of thumb for cartographic representation is to use three to seven categories. Of course, this guideline is violated plenty of times. It really serves as a guideline that is helpful for thinking about cartographic design. For car navigation or mobile phones, three categories is probably the ideal starting point. A land-use map or study of soils will require many more categories, which does make such maps harder to read, but only then can the map meet the professional or scientific needs for which it was created.

Classification is used for quantitative data, including counts, measurements, and calculations. Ordinal, interval, and ratio data can all be classified. Nominal data can also be classified, but only as individual categories or by clusters—for example, land-use types by generic land-use categories, such as urban, mixed, and rural (see Figure 11.14).

Classification is important for cartographic representation. Some people may even claim that its main use in creating choropleth maps makes it at least one of the most important cartographic representation techniques. A choropleth map uses the boundaries of geographic units (e.g., counties, countries, or states) to determine the area represented with a particular shade or color. Since the geographic units are distinct, the cartographic representation makes it easy to compare the characteristics of each unit.

Municipality Areal Units

Number of Inhabitants

24	56	23	17	6	45
21	28	17	21	9	35
27	45	19	39	7	49
25	41	11	27	3	41

Number of Dogs

1	5	9	4	35	10
5	21	7	2	3	14
3	15	4	9	10	12
4	18	5	4	19	17

Dogs per Inhabitant

0.04	0.09	0.39	0.24	5.83	0.22
0.24	0.75	0.41	0.10	0.33	0.40
0.11	0.33	0.21	0.23	1.43	0.24
0.16	0.44	0.45	0.15	6.33	0.41

Aggregated to County Areal Units

Number of Inhabitants

267	174	195

Number of Dogs

72	44	120

Dogs per Inhabitant

0.27	0.25	0.62

FIGURE 11.13. Areal aggregation of geographic units can lead to results that mask important geographical characteristics and relationships.

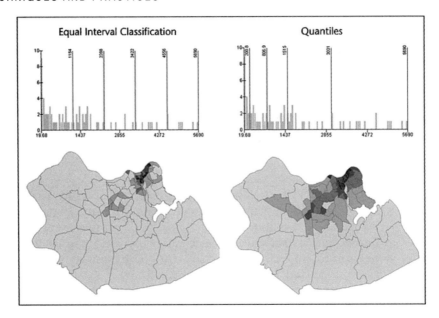

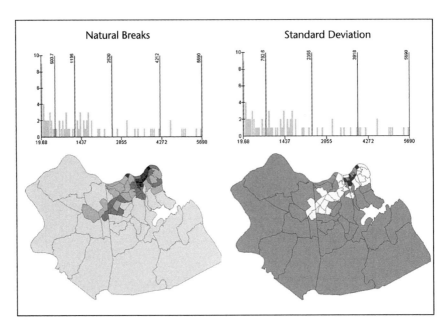

FIGURE 11.14. Examples of four widely used classifications for the same data.

Choropleth maps and visualizations start with defined geographic units, for which data is either collected, or, if the data comes from other data sets, is associated with the same geographic units. These units may be counties, countries, or states, but they can also be an archaeological site, biotope, or watershed. It is of course possible to aggregate the data from one type of geographic unit to a larger unit (e.g., counties to states), but different areal aggregations might result in very different representations, supporting very different conclusions. When creating a choropleth map, it is very important to state which geographic units were used for collecting the data.

In urbanized areas and for environmental applications that rely on administrative or arbitrary boundaries, these issues are of special importance. These boundaries frequently can divide communities or congruent zones—for example, a lake in the middle of a desert. While the water is regionally significant, the lake should only have minute influence on the statistics of a largely desert area. However, if most of the lake ends up in a small town, taken by itself, it may look like the town has rich water resources to itself. In every case, careful selection of the classes is of the utmost importance in crafting a reliable map or cartographic figure.

Types and Applications of Common Classifications

Equal Interval. The equal interval classification divides the total range (the number between the minimum value and the maximum value) into equal parts. This classification is best used for properties of things and events that have an implicit order—for example, the top 30 stores by sales in the state. This classification is valuable for highlighting such an order, but can mask details and deviations.

Quantile. The quantile classification determines classes to ensure that each class has approximately the same number of features. This classification produces results that best show the diversity of a geographic property and can aid in determining relationships, or can lead to questions about possible relationships.

Natural Break (Jenks). Also known as the Jenks classification after the cartographer who developed it, natural breaks use an algorithm to determine where class boundaries should be placed in the total range. The class boundaries should maximize differences and keep similar clusters together. This classification is very valuable for visualizing GI in an insightful manner.

Standard Deviation. The standard deviation classification is different from the other three. It shows graphically how much a geographic unit's property varies from the mean. This is useful for cartographically representing differences in geographic properties from an average value rather than showing exact values.

Symbolization

Things and events require additional abstraction for certain mapping and visualization purposes. The symbols for elements from the geographic representation can be varied in terms of size, shape, value, texture, orientation, and hue. These symbols are used to represent location, direction, distance, movement, function, process, and correlation in an infinite number of ways. The design of cartographic materials draws on long traditions and scientific study of these symbols and their role in cartographic representation and communication.

Pragmatically, size, shape, and hue are perhaps the most significant. Size is important because humans will instinctively understand that a larger dot on a map of the world means a larger city than a smaller dot or that a large runway symbol indicates a large airport. Size shows quantitative differences. Proportional symbol maps use different size symbols as a primary means to communicate quantitative differences (see Figure 11.15). Shape provides clues about qualitative differences. Shape is also important because it gives us an orientation to the actual shape on the ground of the feature and because it helps us to understand geographic relationships. Hue, which refers to what most people refer to commonly as color, is so significant that it requires its own section.

Color, Otherwise Known as Hue

A key component of most maps is color. Humans have an astonishingly broad perception of color, but a surprising number of people face limits in distinguishing color (color-blind people and others with visual or perceptual impairments) and many people will not agree on the name of specific colors. Color is very important for maps and visualization. A good application of color will greatly help with communication.

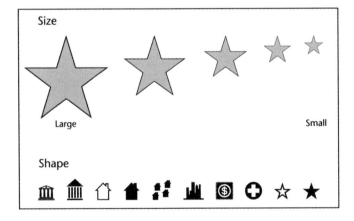

FIGURE 11.15. Size and shape are key visual variables to distinguish quantity and quality in cartographic representation.

The first principle to remember is that the use of color should be discrete, never garish, and never includes overly bright colors that many people (especially the elderly) have difficulties distinguishing. Generally, all people have great difficulties putting colors into a sequence, making color ill-suited for showing quantitative differences. People can better distinguish gray tones, or value, as a sequence from light to dark (see Figure 11.16). The second principle is that colors should align themselves with existing conventions. All people associate blue with water, but other colors may be associated with cultural values and, in some countries, laws require the use of certain colors for certain legal documents and maps.

Color is very complex and differences in displays and printing equipment can lead to astonishing differences. Different standardized color models are generally used for web documents, video, computer screens, and print documents. CMYK (cyan, magenta, yellow, and key—black) is the most common specification for print; RGB (red, green, blue) is widely used for screen displays.

Cartographic Representation Types

Media and Formats

Depending on the media and the format of the map or visualization output, cartographic representation must be adapted. Size and format are important, but the media is a key variable that should not be overlooked.

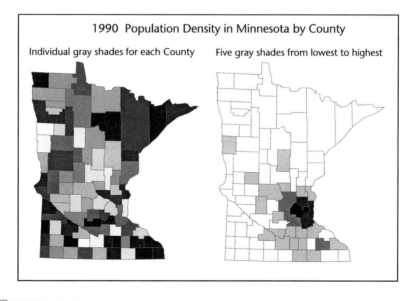

FIGURE 11.16. Ill-advised use of symbols can lead to confusing maps.

IN DEPTH **Polishing the Cartographic Image**

Cartographers might not agree on many points of cartographic design; however, a few suggestions can be a great help if you are just starting to make maps or lack the time to take a course on cartographic design. Following Mark Monmonier, who has written a series of widely acclaimed books on the uses and misuses of cartography, the following rules should guide attempts to improve maps.

1. Be shrewdly selective. Don't show what you'd rather they not see.
2. Frame strategically. Avoid unfavorable juxtaposition and crop the maps and sketches to forestall fears of illness or diminished property values.
3. Accentuate the positive. Choose favorable data and supportive themes for maps.
4. If caught, have a story ready. Computer errors and a stupid drafting technician's use of the wrong labels are plausible excuses.
5. Minimize the negative. If you can't eliminate them entirely, at least don't emphasize features you'd rather have ignored.
6. Dazzle with judicious detail. After all, a detailed map is a technically accurate map, right? Details are useful distractions.
7. Persuade with pap. Try simplistic maps, or maps that camouflage potentially embarrassing details.
8. Distract with historic maps and aerial photographs.
9. Generalize creatively. Filter or enhance details to prove your point.
10. Enchant with elegance, use lots of tree symbols.

Output Types

The output types of maps and visualization usually range from small newspaper, cell phone, or brochure maps to large poster-size maps (see Figure 11.17). Specialized maps can be even larger (e.g., billboard advertisements) or smaller (e.g., postage stamps), and displayed on a variety of materials (e.g., stone carving, metal etching). Most maps and visualizations are either on paper (or a paper-like material) or on some form of computer-controlled display device.

For pragmatic reasons, this discussion focuses on cartographic representation issues only for these most common output types. Paper is still the most common media for cartographic representations. Formats range from small grayscale maps published in newspapers, advertisements, and brochures, to high-resolution large posters of photographic reproduction quality. Small maps need to be simple. If they are complicated, they can become muddled and will communicate poorly. If a small map is printed in grayscale it can easily show quantitative differences, but may become too complicated if many symbol shapes are used.

The output types of visualizations range from small cell phone displays to large computer screens and projectors. However, even if the area of the screen or projected surface is large, the resolution is still very low

▌FIGURE 11.17. GIS digitization, viewing, and plotting equipment.

in comparison to that of printed maps. The lower resolution can be compensated for by offering capabilities to zoom and pan in the visualization. This makes it possible to interactively move around a visualization based on a user's interest. While this is an essential feature for small displays—cell phones and similar devices—it is also necessary for larger displays of visualizations because of their lower resolution.

Types of Output Equipment

Output equipment for cartographic visualizations can be separated into two groups. Printing equipment produces output on paper or paper-like material. Visualizations are made with displays.

More and more printing equipment has become available to produce color maps, but the costs remain comparatively high. A color ink-jet printer may cost less than $50, but the ink cartridge for 30 full-color maps may cost another $30. Printing costs of $1.00 per unit for a small color map is very expensive. For large series of maps, such costs can be prohibitive. Grayscale maps, printed with laser printers, ink-jets, or photocopied are far cheaper, remain common, and will likely be widely used far into the future.

Depending on the resolution of the printing equipment, care should be taken making grayscale maps. If the printing equipment has a very low resolution (e.g., a fax machine), the clarity of the high-resolution grayscale map may degrade into a collection of unsightly ink spots. Issues with printing equipment resolution also play a role in making large-format maps. Because of differences in printing technologies, maps that portray well when previewed on a computer screen may appear much worse when printed on paper. If a substantial map is being produced, it is wise to check the quality of the final map with a sample before commencing a large print-run.

High-end color printing equipment supports resolutions of over 1,000 dots per inch (dpi) and brings great detail to printing. If the color printing equipment supports a much lower resolution (e.g., the 300 dpi of most large-format color ink-jets), then the print quality difference will be noticeable, but is usually not enough to distract from engineering or planning applications (see Figure 11.17).

Communication Goals

Clearly the choices of media, format, and output depend on costs and available resources, but the communication goals remain the most important. Whenever possible, the communication goals of a map or visualization should be taken into account at the beginning of working on the geographic representation or cartographic representation.

Depending on media and format, the same communication goals may require different geographic representations. Advertising for a new clothing store in a monthly magazine will need a different map than the store location map on a website or available for cell phone users on request. There is rarely a need to create three different geographic representations to support these different communication goals, but it is beneficial if the geographic representation can consider the goals and create GI to support them all.

The early work on a geographic representation dovetails well with work on a cartographic representation that supports different communication goals. Generalization offers some flexibility for using GI in different ways, but additional effort is usually required to assure that the cartographic representation supports the communication goals. This usually involves the use of different symbology for different output media and formats.

Distortions

Different output media and formats have different levels of distortion. This can be very important if the maps or visualizations are being used for technical or legal purposes, but the distortions resulting from generalization may be far greater.

Most of all, the process of cartographic representation can produce changes in the position of objects. Road geometry is moved to clearly show the connectivity and to remove conflicts with other feature symbols. These changes may not be indicated, and if the altered GI is transferred to others without an indication of the changes, its use for other purposes may lead to erroneous results. It is also possible that attribute changes, usually as a result of generalizing GI, can later lead to grave errors and limited positional and attribute accuracy.

Types of Presentations

Creativity knows no bounds and cartographic presentations evidence wonderful creativity in an endless gamut. This section of this chapter presents a mere selection with a pragmatic emphasis on the most common types. In practice, many of these types are combined into hybrids, but the following suggest the range of possibilities. Depending on one's experience, training, discipline, and institution, other approaches will be encountered.

Topographic

The universal map was developed for military and civilian use in the 16th century by European countries, but in the 19th century it took on a form that made it the most significant kind of mapping for governance, the military, and colonization. The military often commissioned or heavily influenced topographic maps. Significant details and secret information would be edited out of civilian versions of maps. In some countries, distortions were introduced to assure that even generic information could never be used to locate other things or events. Ideally at a scale of 1:25,000, or even 1:10,000 for very detailed maps, the high costs of mapping often led to topographic maps being produced at scales of 1:50,000 or 1:100,000 (see Figure 11.18).

Cadastral

Cadastral maps show land ownership, rights to land access and use, and obligations. In most countries they are best known for their use in taxation, but this only becomes a significant revenue source in a few countries of the

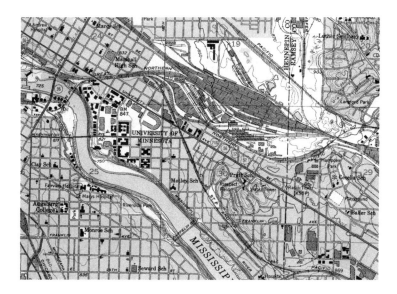

FIGURE 11.18. Portion of the 1947–1951 USGS topographic map from Minneapolis, Minnesota. Maps such as these are printed using very high resolution printers.

world, especially the United States. The cadastral map's significance comes mainly from its role in governance and in land development and speculation. A government wishing to pursue developments needs cadastral mapping in order to assure it knows where it can develop and who should be involved. In times of upheaval, the cadastral map has also been an aid for taking land away from groups and individuals and turning it over to others (see Figure 11.19).

Thematic

Thematic maps, both qualitative and quantitative, are perhaps the most widely used form of maps. They make it possible to show things and events that people can otherwise not see nor experience—for example, 2014 birthrates in counties, or population per square mile in all 50 U.S. states (see Plate 10). The major strength of this form of cartographic presentation is that it enables easy comparison. The birthrate, other demographic statistics, or environmental statistics can be presented in a tabular form, but the use of a cartographic presentation helps with comparisons by making the geographical relationships plain.

Choropleth

A special type of thematic maps, choropleth maps use a graphic variable (usually hue or value) to show quantitative differences. They are easily prone

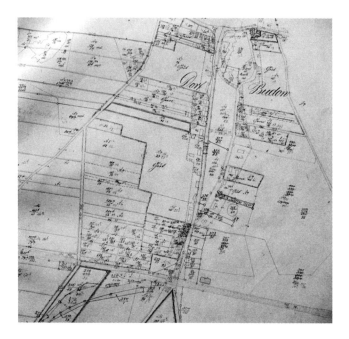

FIGURE 11.19. Portion of the 1865 cadastre of Budow (now Budowo). Cadastre maps are some of the oldest types of maps known.

IN DEPTH Dasymetric Mapping

John K. Wright developed this technique for mapping that merges geographic and cartographic representation. Recognizing that choropleth maps fail to distinguish between subareas of an aggregation unit (e.g., urban and rural areas in counties), they will show amount of grain grown per hectare for urban areas where no corn or wheat is grown, and show average number of people in a family for rural areas where no people live.

Wright's solution subdivided the aggregation unit into subareas that were assigned weights based on associations between the land-use type of that subarea and the property to be mapped. The weights add up to 1 for the entire aggregation unit.

The dasymetric technique is a far more accurate way of representing GI than choropleth maps and can be relatively easily done using GIS overlay operations and some calculations. See Plate 11 for an example of the difference between a choropleth and a dasymetric representation.

to misunderstanding because they fail to distinguish differences between subareas of the geographic units shown. For example, a map showing birthrates in the 50 U.S. states fails to distinguish urban and rural differences. Choropleth maps are also erroneously used to show the number of a particular property for an area—for example, population, cars, or voters. However, the area of the geographic unit (e.g., state or county) plays a direct role on how much of the property can be found in an area. No matter how urban an area is, the number of religious buildings stands in relationship to the size of the area. The easiest way to avoid this error is to always use choropleth mapping for properties of a constant unit (e.g., an acre or hectare). This approach was further developed as dasymetric mapping (see Figure 11.20).

Charts

Charts are still widely used for naval or aerial navigation, but they can be used in conjunction with maps to provide geographic orientation. A variety of charts at different scales is produced by national and international bodies charged with ensuring the safety of navigation. Using up-to-date geographic information and charts at an appropriate scale for an activity remains of great importance.

Cartograms

Cartograms, both contiguous and noncontiguous, show quantitative difference by altering the size of the geographic units according to the relative proportion of the geographic unit's property. One type of cartogram stretches all boundaries to create an image that roughly corresponds to the original, even if greatly distorted. The second type simply scales the boundaries of each unit according to its rank with other units.

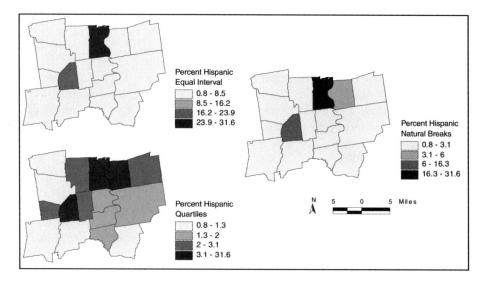

FIGURE 11.20. Examples of choropleth maps of the same data using different class interval methods.

Dot-Density Maps

A dot-density map or visualization is used to show the quantity of a property by placing a dot in a geographic unit for each incident of a property (or multiple of incidents). The dots can either be placed to correspond to the position of the incidents, or randomly placed within each geographic unit. The former provides an excellent understanding of the property's geographic distribution, but must be prepared by hand or with specialized software (see Figure 11.21).

Summary

Principles of cartographic representation include scale, generalization, classifications, media formats, and cartographic presentation types. Geographic representation considers how to abstract things and events from observations and measurements of the world. Cartographic representation deals with the creation of visual techniques and forms following cartographic principles. Depending on the means and modes of communication, different cartographic representations could be required. A classroom wall map of Europe or North America will show things and events differently than a small map of Europe or North America for a news broadcast. The design of maps and visualizations is essential to successful geographic communication.

Key concerns for cartographic representation center on the design of maps and other material that reliably abstract from and with geographic

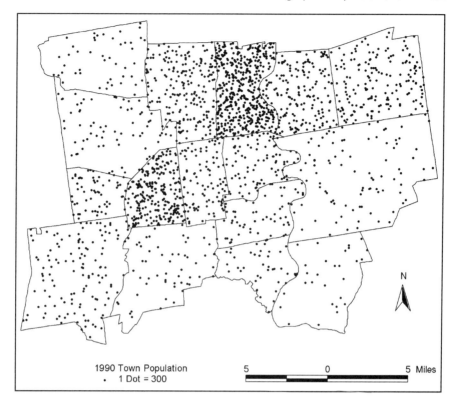

1990 Town Population
. 1 Dot = 300

5   0   5 Miles

N

FIGURE 11.21. A dot-density map of population density.

representations. Cartographic representations are in many ways inseparable from geographic representation. High-quality communication often conceives of both simultaneously, although the process of creating GI frequently is separate. In the end, a cartographic representation should retain the quality of the GI. Otherwise substantial errors and distortions in the creation of maps and visualizations and their use can occur.

REVIEW QUESTIONS

1. What is the ideal number of classes to show on a map?

2. What does the larger size of a symbol generally imply?

3. When is cartographic animation generally useful?

4. Why is it important to know the type of output equipment when considering cartographic representation issues?

5. What makes cartographic symbols iconic?

6. What are five suggested map elements?

7. What should symbolization show?

8. When is an equal interval classification useful?

9. How is a histogram useful?

10. Why is the Jenk's classification widely used?

ANSWERS

1. What is the ideal number of classes to show on a map?

   According to cartographic research, between three and seven classes is ideal.

2. What does the larger size of a symbol generally imply?

   Size should be used only to show ordinal distinctions.

3. When is cartographic animation generally useful?

   Cartographic animation is generally helpful for showing events when only discrete GI is available.

4. Why is it important to know the type of output equipment when considering cartographic representation issues?

   Output equipment points to specific choices which affect the resulting cartographic presentation.

5. What makes cartographic symbols iconic?

   The use of commonplace symbols to achieve strong, heterogeneous semiotic references makes cartographic symbols iconic.

6. What are five suggested map elements?

   The five most commonly suggested map elements are scale indicators, title, author, date, and north arrow. Explanation and contact information can additionally be considered.

7. What should symbolization show?

   Symbolization should show enough reference to relevant and meaningful things and events without unnecessary exaggeration and conflict with other symbols. The whole map should balance cartographic representation needs with the underlying geographic representation.

8. When is an equal interval classification useful?

   The equal interval classification is suited for ordinal, interval, or ratio ranges of attribute values that are evenly distributed.

9. How is a histogram useful?

   A histogram is useful for graphically showing the range of attribute values.

10. Why is the Jenk's classification widely used?

Based on statistical techniques to identify clusters in data, the Jenk's classification minimizes variance within clusters. The homogeneous classes usually offer a better cartographic representation with wide-ranging attribute values.

Chapter Readings

Cromley, E. K., & McLafferty, S. L. (2011). *GIS and public health* (2nd ed.). New York: Guilford Press.

Kraak, M. J., & Ormeling, F. J. (1996). *Cartography: Visualization of spatial data*. Harlow, Essex, UK: Longman.

MacEachren, A. M. (1994). *Some truth with maps: A primer on symbolization and design*. Washington, DC: American Association of Geographers.

MacEachren, A. M. (1995). *How maps work: Representation, visualization, design*. New York: Guilford Press.

McMaster, R., & Shea, K. S. (1992). *Generalization in digital cartography*. Washington, DC: American Association of Geographers.

Monmonier, M. (1991). *How to lie with maps*. Chicago: University of Chicago Press.

Robinson, A. H. (1984). *Elements of cartography*. New York: Wiley.

Slocum, T. A., McMaster, R. B., Kessler, F. C., & Howard, H. H. (2008). *Thematic cartography and geovisualization*. New York: Prentice Hall.

Web Resources

🕑 ColorBrewer offers an excellent resource on uses of color in cartography online at *http://colorbrewer2.org*.

🕑 For information on geovisualization research and concepts, see *www.geovista. psu.edu/sites/icavis/icavis/ICAvis_overview(1).html*.

🕑 For practical guidelines for U.S. NPS mapmaking (Chapter: Mapmaking for Parklands), see *www.nps.gov/jeff/historyculture/mapmaking.htm*.

🕑 The practical mapmaking guidelines developed by the U.S. Forest Service are available online at *http://purl.access.gpo.gov/GPO/LPS17532*.

🕑 Some very interesting information graphics produced with and based on maps can be seen at *http://fathom.info*.

🕑 The folks at Y-Worlds reflect on mapmaking: *http://daily.yworlds. com/2012/12/31/map-makers-2/*.

🕑 Many Eyes by IBM offers rich examples and possibilities to make maps and visualizations; see *www-958.ibm.com/software/analytics/manyeyes*.

EXTENDED EXERCISE

Map Measurements

Overview

In this exercise you will learn techniques for measuring distance and area on maps. An important part of making measurements is assessing their accuracy.

Objectives

Of the different ways to measure distance and area from maps, the most common techniques use specialized instruments (an opisometer or a planimeter) or straight-edges and grids. Accurate measures are important. A variety of processes can negatively impact accuracy. The scale of the map is one such issue. Generalization may also have impacts. Warping and stretching of paper, or imprecision in map production, can also have more important effects. In this exercise you will learn how to make measurements from maps and gain some insight into the results of generalization and other factors on the accuracy of map measurements.

Steps and Questions

For steps 1a and 1b you will need a map sheet at the scale of 1:250,000 and a map sheet at the scale of 1:24,000 that cover the same area. If your map sheet is out of the Public Land Survey (PLS) in the United States, skip the questions that ask for township and range information.

Step 1a—Get a 1:250,000 Topographic Sheet

Find a 1:250,000 topographic sheet for an area that overlaps with the 1:24,000 topographic map sheet you get in step 1b. Also make sure that the 1:250,000 sheet includes all the features you will need to locate in Step 1b. Identify a feature (small building, intersection, pond, etc.) on both map sheets that you will use for calculating distances.

What is the name and/or type of your feature? _____

What are its UTM coordinates? _____

With your straightedge, or paper-strip, measure the distance of this feature to three other features:

Distance (in meters) to nearest road intersection _____

Distance (in meters) to highest elevation point _____

Distance (in meters) to nearest township/range corner _____

Step 1b—Get a 1:24,000 Topographic Sheet

You should measure the location of the same feature you worked with first in the Step 1a feature:

What is the name and or type of your feature? _____

What are its UTM coordinates? _____

With your straightedge or paper-strip, measure the distance of this feature to three other features:

Distance (in meters) to nearest road intersection _____

Distance (in meters) to highest elevation point _____

Distance (in meters) to nearest township/range corner _____

1. How large (in page units) is the feature you measured from? How will its size, especially in comparison to the distances you measured, affect the accuracy of your measurements? Will they be more significant than generalization effects?

2. Do you recognize any generalization effects in the smaller-scale map? Considering all five types of operations, which ones? Describe or sketch examples.

Step 2—Measuring Area

A number of grids are available for you to use in this step. Take one of the 1:24,000 Scale Grids and one 1:250,000 Scale Half Mile Grid sheet. Each cell of the 1:24,000 Scale Grid is 660 × 660 feet, or 435,600 square feet (equivalent to 10 acres). For our purposes, on the 1:250,000 Scale Half Mile Grid, each cell (approximately 1/8″) is equal to the same length and area.

Determine the areas of some topographic features using the two following techniques. If a portion of a feature fully fills a cell, it covers an area of 435,600 square feet. Partial cells can be treated following technique A or B. You will use both techniques so be sure you understand them:

A. For each cell partially occupied by the feature, estimate the proportion of the cell that is occupied. Compute the sum of these values, then add the sum to the area of the completely occupied cells.
B. Count the number of partial cells and multiply them by the area of an individual cell. Then divide by 2. Next add this sum to the area of the completely occupied cells.

To measure area, position the lower-left corner of your area grid at the lower-left edge of the feature you will measure. Determine the areas according to technique A or B.

Determine the size of the following areas in square feet on your 1:24,000 topographic quadrangle. If you use the "English Area Grid," you should use the printed conversion factor to determine the area in acres. Otherwise, convert the number of cells using your value from Technique A or B.

The PLSS section your feature is located in:

Technique A _____

Technique B _____

A nearby lake or pond:

    Technique A _____

    Technique B _____

A city block or town extent:

    Technique A _____

    Technique B _____

3. Are there any differences between technique A and B? How do you explain them? Which technique do you think is preferable? Make sure to consider the different sizes and shapes of features you measured.

Step 3—Compare Scale Effects on Area Measurements

Find an areal feature (at least 2″ × 2″) on your 1:24,000 scale topographic map and measure it using techniques A and B. Make sure you can find it on your 1:250,000 scale map:

    Feature name and type _____

    Area using:

        Technique A _____

        Technique B _____

Find the same feature on your 1:24,000 scale map and measure its area using techniques A and B:

    Area using:

        Technique A _____

        Technique B _____

4. Is the feature larger or smaller on the 1:250,000 scale map? What generalization effects do you find? Explain the differences you detect with specific examples (e.g., the road defining the park's northern border is greatly simplified in the 1:250,000 scale map). Make a drawing if it helps.

	Point	Line	Area
Aggregation	X	X	X
Displacement	X	X	X
Enhancement		X	X
Selection	X	X	X
Simplification		X	X
Smoothing		X	X

CHAPTER 12

Map Cultures, Misuses, and GI

GI and maps are in their current form because of how they developed, and that history is related to cultural issues and values. The visual representation of GI and maps gives them power. Though GI and maps have a utilitarian component, they are not a simple device. Rather, they are a vehicle for complex symbolism in which present images are built on the past, and the shorthand of their symbolism can mask ideological and material concerns. GI and maps can communicate in different ways and at different levels to the people they reach. Around the world we find very different global traditions of mapping. And while geographic and cartographic representation draw from a multitude of symbol systems, many of these follow conventions that are often implicit. GI and maps guide not only *understanding* but also have the potential to *influence* what people know. An individual only ever sees a small portion of the world, but maps give us the means to engage with and even manage more of the world than we can see. Seen from different cultures, the same map can seem wrong or distorted because of different cultural values and experience. Some might even say that misuse of GI and maps is built in to exchange and conflict among cultures. Developing an understanding of the cultures of GI and maps helps develop a better sense of how GI and maps give meaning to geographies, how people manipulate GI and maps to influence others, and how the misuse of maps can lead to biases in how we understand the world or parts of it.

The goal of this chapter is to consider the cultural problems inherent in geographic and cartographic representation in Western civilization. These include accuracy issues discussed in previous chapters. Perhaps the question goes all the way to why GI and maps are created in the first place. These activities are very costly.

Cultures of Maps

Culture is a word with many meanings. To discuss the culture of GI and maps, we should distinguish between "national culture," "indigenous culture," and "disciplinary culture." **National culture** refers to the overarching set of values, beliefs, and implicit understanding commonly held by the majority of people living in a particular nation. **Indigenous culture** corresponds to the beliefs and knowledge of people who lived in an area before it was colonized and to their heirs. **Disciplinary culture** indicates the specific set of values, beliefs, and implicit understanding widely used and accepted by a particular profession or discipline—for example, sanitary engineers and city planners in an urban realm, or conservationists and consulting engineers for environmental issues. The beliefs of individuals change and cultures blend. This mitigates oppositions among cultures. Many people have multiple professional responsibilities and have different professional and life experiences

---

IN DEPTH **Different Ways to Represent Geographic Knowledge**

Textbooks like this book focus on the creation and use of GI and maps for the representation of geographic knowledge, but we certainly should at the bare minimum mention other forms of representation: text, pictures, drawings, and verbal descriptions. They are often used in conjunction to expand or clarify the graphics of a map.

Text is perhaps the most common form of geographic knowledge representation because it is easier to create than GI and maps, but is obviously limited. However, as we all know from experiences giving and getting directions, text is often far more convenient and easier to use than a map drawn by a friend or colleague. For particular purposes, it may even be easier to use than a generic highway or city map.

Pictures even can look like maps, but often simply use one or two conventions to help readers understand them as maps with geographical relationships. Maps used in advertising or at tourist destinations often are really just pictures that may add a legend and standardized symbols.

Drawings are a third type and are very common, but usually limited to communicating geographic knowledge related to a particular purpose—for example, comparing the size of two shops, streets, or countries. Drawings may loosely rely on the conventions of mapping, but are rarely consistent due to the limited scope of their intended use.

Indigenous cultures often rely on verbal descriptions, not just for communicating where things are and how to get to places, but to share complex and important stories about the culture's creation and meanings associated with places. These meanings can reflect generations of experiences, replacing scientific observations and measurements shown on a map with a deep lore and understanding of place.

that impact their own cultural understanding. Regardless of the variety of cultures and difficulties of pinning down the exact influence of any culture, cultures always mediate how we understand and influence geographic and cartographic representation in GI and maps.

Civilizations and Maps

Before looking at more specific influences of culture on GI and maps, we might want to start by taking a step back from specific cultural issues of geographic and cartographic representation and exploring the broader relationships between civilizations and maps. As far as archaeologists can tell, all human cultures have had some form of representing geographic things, events, and the relationships among them. These have taken the form of prehistoric cave paintings showing spatially positioned animals with hunters attacking them, sticks bound together to depict a nautical map, a story carved in rock that shows a spatial relationship between a tribe and the universe that encircles it, and all kinds of objects, art, and writing from the past that people today would see as being "maplike."

European

From the Iron Age on, European cultures and cultures in the Middle East—or plain Western civilization, if we want to avoid dealing with the complex differences and conflicts among the different national cultures of Western Europe and many colonized countries—put maps and related graphical imagery at the heart of how many people came to understand the world. Maps in the past complemented experience in much the same way as today by depicting things and events beyond most people's experience. For example, the Romans relied on maps to show the extent of their empire. The maps became iconic, as an image of the geographic extent of the empire. Before the Romans other cultures also used maps. Sometimes maps took on more of an iconic role, and sometimes maps were simply used for practical tasks.

The Romans developed maps in both their iconic and practical roles further. Iconic maps served, as Denis Cosgrove writes, as a way for individuals to understand their place in an empire and to associate their calling and position in life with both the mundane events of day-to-day life and the divine, symbolized by the emperor for most of Roman history. Maps were integral to the Roman cosmology. On the practical side, Romans not only advanced but solidly established the basic principles of cadastral maps (see Chapters 5 and 13), which would form the basis for modern arrangements of land ownership and subdivision hundreds of years later. The itineraries used by the Romans for commerce became the basis for showing travel times between cities. Maps showing the length of connections between places based on time are still important for transportation, especially subway maps that show relationships between stations based on connections, not physical distance.

From the late Middle Ages (around 1400 A.D.) on, maps took on more practical functions for military uses but also for cadastres and increasingly

> ## IN DEPTH   Placenames and Conflicts
>
> The names given to a place may change over time and may reflect deeper changes in local society and culture. However, usually only one name is presented on a map. Most maps indicate the capital of Italy and location of Vatican City with the Italian name, Roma, but some may replace it with their own language's name—for example, in English, Rome; in Polish, Rzym; in German, Rom; in Japanese, マロー; and in Chinese, 羅馬. This may be common practice in a country's schools and in the general media, but a globalizing world has meant that people need to be more attentive to naming practices.
>
> This has long been the case in multicultural states. For example, in Switzerland, where four languages are nationally recognized, cities in zones with influences from different cultures may often be identified on maps with two names—for example, Delsburg and Delemont or Neuenburg and Neuchâtel. Cities clearly in one language area will usually be shown with only one name, although exceptions occur. These exceptions can be laden with conflict and be irresolvable. For example, the large lake that Geneva lies on is known there as Lake Geneva, but along the majority of the shoreline in Switzerland and in neighboring France, the lake is known by the name of Lac Leman, which derives from the Roman name of the lake, Lac Lemanus. To this day, Geneveans and the international community use their name, while others use Lac Leman.
>
> Conflicts over naming occur around the world. A significant geographic naming conflict is between Japan and Korea over the name of the sea between the two countries. The sea was known variously as the "Korean Sea" or the "Japanese Sea" for most of the modern period. When Japan began expanding into Manchuria and Korea in the 20th century, the name became the now commonplace "Sea of Japan." For some time now Koreans have been seeking to have the name changed officially by the United Nations (UN) to reflect historical practices, but although their claim has been acknowledged, its recognition and adoption by the UN has been slow in coming.

for the visualization of statistical information used for social and environmental decision making. Jumping quickly to our information age, it's easy to find examples every day of the iconic role of maps. Just consider many company logos and advertisements. Maps (and later GI) have become part of the increasingly specialized and bureaucratized social activities that require detailed understanding of the world and the ability to share this understanding with others. The military has played a key part in map-development and surveying activities, although commercial interests also have produced GI and maps or adapted what the military has generated.

Indigenous

Native, or indigenous, cultures around the world have used forms of geographic representation that should also be considered as maps. Even though they may be sharply different from what people in Western-influenced cultures have come to understand as maps, they are material devices used to communicate things and events in their geographic relationships. An example of this type of map is the stick charts used by Pacific South Sea islanders as training devices and navigation aids for people traveling across the vast expanses of the Pacific Ocean (see Figure 12.1).

Other indigenous cultures had other forms for communicating geographic ideas, often connecting spiritual and physical geographies. The failure to recognize indigenous cultural geographic representations has been an ongoing source of conflict in many areas of the world. Rundstrom, along with many others, points out that the assumption that a European-based representation of the world in a topographic map contains just the "facts" represented by naming can lead to the inclusion of sites and artifacts that hold special spiritual and cultural significance for indigenous groups. As these sites and artifacts become attractions for people from other cultures, their mapping has led to people coming and destroying the sites by removing items or disrespecting the indigenous culture's activities.

Cultural Forces within "Disciplines"

A very different kind of culture also has a great effect on GI and map creators and users and comes from within "disciplines" themselves, even though it may be hard to identify specific values, beliefs, and so on. Because of the legal significance of the maps they often create, surveyors have concepts of geographic

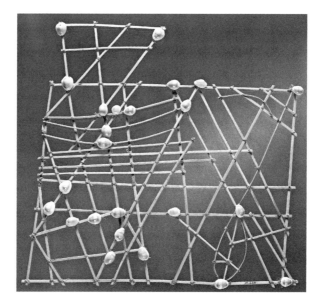

FIGURE 12.1. Polynesian stick-and-shell map. Many civilizations around the world have used some form of mapping to record geographic relations involving things and events. Indigenous groups developed very complex approaches.

and cartographic representation that are different from those held by other geography-oriented professionals, for example, planners. The accuracy with which the location of a road center line is determined and mapped differs if, for example, the map is merely showing the highways in the state versus whether the map is being used to formalize a right-of-way, possibly leading to seizure of private property for a needed highway improvement. Scale and resolution requirements for geographic objects will often be different among different disciplines; how features and events are selected for inclusion in the discipline's geographic representation also will differ significantly.

City and state planners, and planners at other levels of government, will often interact with or employ surveyors. Yet the two groups have somewhat different cultures. The emphasis of the planners may be on communicating their concerns, their "plans," to the communities affected by their plans. Building needs, transportation, the rational use of infrastructure and resources, avoidance of undue hazards, and concern for environmental quality require flexibility. In a way, the maps that planners produce are always evolving, being redrawn. The culture of surveyors, on the other hand, depends on great accuracy (the flexibility of the planner would not be prized), recording the location of property lines, and such. Geographic and cartographic representation in the culture of surveyors often has a legal component and follows professional regulations and long-established practices (see Figure 12.2). Indeed, land features as characterized by surveyors

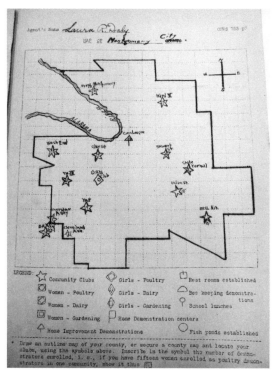

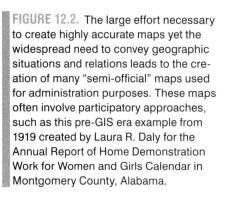

FIGURE 12.2. The large effort necessary to create highly accurate maps yet the widespread need to convey geographic situations and relations leads to the creation of many "semi-official" maps used for administration purposes. These maps often involve participatory approaches, such as this pre-GIS era example from 1919 created by Laura R. Daly for the Annual Report of Home Demonstration Work for Women and Girls Calendar in Montgomery County, Alabama.

IN DEPTH Crises of Representation:
Crises for GI and Maps?

A number of social philosophers have criticized the representational culture of Western modernity since the late 19th century. Their impact has been substantial and broad. For GI and cartography, their endeavors to understand the activities of representation and construction of GI and maps have led to a number of contributions, which have begun more and more to influence scientific and professional cultures. In his book *A History of Spaces*, John Pickles engages these issues by developing three crises of representation that I interpret here in the context of broader crises of representation.

The first of these crises relates to the assumption of objectivism in GI and maps. Many GI- and mapmakers and users presume that their GI or maps approximate the real world as a correspondence or true relationship between symbols on a map or points, lines, and areas with a single "reality." For them, the making of GI and maps involves the straightforward collection, preparation, transmission, and reception of information. In other words, a good map is one in which the information intended for communication by the maker corresponds to the information received by the map user. The map, in objectivism, is then an accurate representation of the real world. Many people, of course, have come to realize that the accuracy of GI or maps has as much to do with what people are trained to see and measure. Out of the crisis of objectivism, people recognize that GI and maps are as bound by conventions and ideologies as they are by the science they deploy or follow.

The second of the crises relates to the assumption that what GI and maps represent is natural. Clearly a corollary of the first crisis, many people who wish to believe that maps correspond to what can be found in nature have ended up finding out that any such correspondence results from the role of maps in first creating what is to be found—for example, wetlands or low-density housing. The boundary between the scientific and the political roles of maps is very fuzzy, and for some any attempt to draw a boundary is inadequate. The intentions of GI and mapmakers and users matter. For David Turnbull, this crisis culminates in the recognition that maps precede territories, creating the spaces they map and then claiming them to be natural.

The third crisis is that of subjectivism. The knowledge represented in a map is always connected to forms of social interest. Biases in apparently "objective" GI and maps result. This crisis involves the recognition that we always must consider the interests of GI and map users and makers when considering the cultures and roles of GI and maps in any society.

and surveying is a body of work that can go back hundreds, even thousands, of years.

The distinctions between disciplinary maps in an area are often largely cultural. Yet the people who create and use GI and maps are often unaware of how the GI or the map was influenced by cultural forces. One of the underlying reasons GI and maps are so powerful is that if there are distortions in a map or in the area it depicts, one must know a great deal about the area to ferret out those distortions. The seminal work by Denis Wood demonstrates that cultural values connected to ideology guide GI and maps to highlight a desired understanding of geographic features and subtly erase others in the representation. In one example, Wood analyzed a highway map of North Carolina in which the mapmakers had positioned explanatory text over a poor area of the state, thus obscuring it, and made several other such choices.

Since we will all only ever physically see or experience a small portion of the world, GI and maps offer us the means to extend our "sight" in immeasurable ways. Covering portions of a state on a map subtly but significantly changes what we can know about the state from that map. In effect, that covered area disappears for the map reader. Creators of GI and maps hold a unique position in harnessing the power and authority of a map and creating graphics with agendas. The responsibility to explain both outspoken and vested interests of the GI or map lies in their hands.

Representing Other Cultures: Participatory GIS

A significant response to the selectivity of GI and maps arose in the late 20th century and has come to be known as participatory or public participatory GIS (PGIS). Multiple strands of development and also multiple emphases of activities can be grouped under the term PGIS. Some of the first prototypes of GIS-based analysis were developed by Ian McHarg in the 1960s with an explicit participatory concept in mind, but the concept of PGIS arose mainly from broad academic discussions of GIS and society in the 1990s. PGIS grew out of attempts to move beyond social theoretical critiques to promote the development of GIS that could practically support the needs of communities.

The integration of GIS in a community can take a variety of forms. One of the most common is the development of government-supported provision of equipment, experts, and training for local community groups (see Figure 12.3). Significant also is the development of sense-of-community GIS through local user groups, who work independently of any government group. These grassroots GIS may become very significant in the development of capabilities for local groups to develop and use existing GI and help foster a community sense of place, which has had important political, social, and economic consequences.

A key challenge for PGIS is the explicit integration of local knowledge. This enables powerful and insightful analyses, but runs the risk of making information available in forms that other people and groups may appropriate without accounting for the people who have developed and nurtured

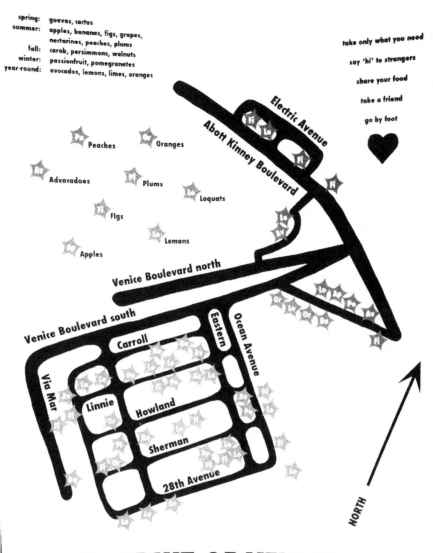

spring: guavas, cactus
summer: apples, bananas, figs, grapes,
      nectarines, peaches, plums
fall: carob, persimmons, walnuts
winter: passionfruit, pomegranates
year-round: avocados, lemons, limes, oranges

take only what you need

say 'hi' to strangers

share your food

take a friend

go by foot

Electric Avenue

Abott Kinney Boulevard

Peaches

Oranges

Advacadoes

Plums

Loquats

Figs

Lemons

Apples

Venice Boulevard north

Venice Boulevard south

Carroll

Eastern

Ocean Avenue

Via Mar

Linnie

Howland

Sherman

28th Avenue

NORTH

FALLEN FRUIT OF VENICE BEACH

this map is a template for free use.   there is no copyright.   learn your fruits!   more information at fallenfruit.org.

FIGURE 12.3. Simple maps, just showing the topological relations in an area, can be a simple and effective means of communication to increase participation. Fallen Fruit maps in the United States and from similar projects around the world use these techniques to help people find valuable food resources that might otherwise be lost to the community.

local knowledge, often over generations. For example, if locals identify a site used for religious rituals, this can enable unscrupulous relic traders to find the site and disturb or possibly destroy it, as happened when the tomb raiders destroyed many Egyptian tombs.

PGIS, once implemented, can help local groups take on government functions. This can be a two-edged sword. On the one side, the use of GI and maps can be empowering and help communities cope with changes and make plans for the future. On the other hand, using GI and maps often requires training experts who end up limiting access to these resources, in effect creating a new elite. The problems of incorporating GI and maps that come from people whose training is not specifically from cartography or geography can be especially challenging. PGIS is a frequently used term encompassing participatory mapping, public participation GIS, counter-mapping, and qualitative mapping. These signal the diversity of different strategies that are currently challenging, mitigating, or opposing authoritative mapping (see Figure 12.4 for an example). Too often, though, these have

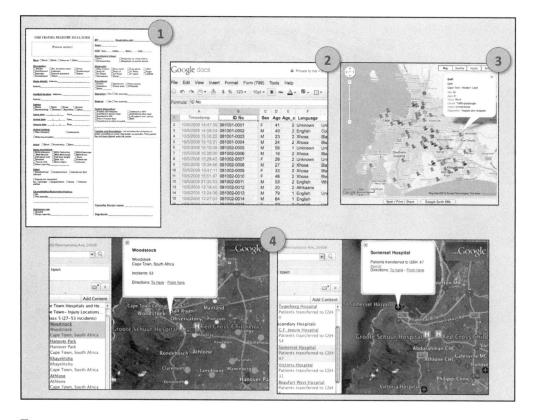

FIGURE 12.4. The preparation of data for participatory mapping can involve multiple steps and complex coordination. In this case, data was collected by local groups to help doctors and emergency response understand the location of various accidents and injury types in Johannesburg, South Africa.

neglected or even willfully erased indigenous and minority geographies and cultural values.

GI and Map Misuses

"All maps lie" is the statement made by the cartographer Mark Monmonier, drawing on the idea behind the classic book *How to Lie with Statistics* and applied to cartography. Many cartographers would like to suggest that "lying" is nothing more than the simple "white lies" that mask the important and complicated work done by cartographers in making maps (see Figure 12.5). As we have read earlier in the chapter, however, there are substantial differences between the distortions necessary to improve cartographic communication and the distortions that erase and obfuscate relevant things and events from the cartographic representation. We need to be able to distinguish between acceptable and unacceptable cartographic representations. In other words, we need to know when GI and maps are misused, or, in the extreme case, when GI or maps are propaganda (see Figure 12.6).

When Is Distortion Propaganda?

Taking the definition that **propaganda** is information manipulated to fit particular ideological, political, or social goals with malicious intent, then the distortion of any cartographic representation based on a person's or group's ideological, political, or social goals makes for propagandistic GI and maps. Examples are all too commonplace. Certainly, examples from Nazi Germany and the cold war are blatant, but other examples can be found in everyday advertising and political publications. Following Denis

---

IN DEPTH **What Is Propaganda?**

Propaganda, for most people, is the systematic manipulation of information to fit particular ideological, political, or social goals.

Thinking about the role and creation of maps in propaganda, we might want to start out with the insight that propaganda maps rely on distortions. Some are blatant and some are subtle, but these distortions, when critically examined, manipulate the GI or map to create a representation that sets out to convince that it, regardless of distortions, is the correct representation of that aspect of the world.

The quiet persuasiveness of GI and cartography is what makes it such a powerful form of deceitful manipulation. When packaged as effective communication, many people may not be able to assess the distortions in GI and maps. Ultimately, the propaganda of GI and maps has little to do with scientific aspects of geographic and cartographic representation, but are based in particular social and political interests advanced through the map.

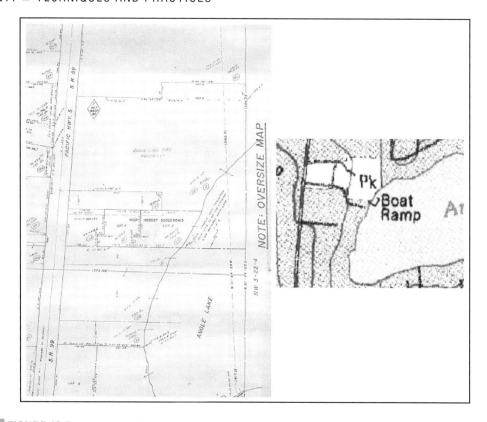

FIGURE 12.5. Angle Lake, Washington, in assessors and USGS topographic map representations. The differences in the maps reflect distinct functional and disciplinary cultural requirements.

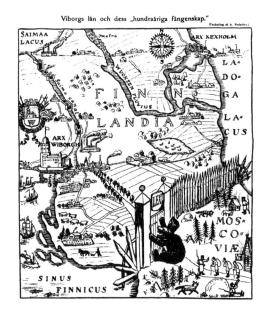

FIGURE 12.6. A complex allegory and elements of a map from the early 20th century showing the Russian bear and the Russian army held back by a stout barricade from the culture, farms, and industry of Finland. Many people would call this map an example of propaganda.

IN DEPTH Errors in the GPS Data? Just That, or Examples of Interference?

Many people around the world rely on GPS, or increasingly GNSS, daily for determining their location, finding their way, or getting information about the location of other activities, things, and events. While GPS is likely to be a major source of data into the future, it is important to note that while most GPS errors are just unintentional consequences of natural and man-made phenomena, because of the low power of the GPS signals, it is actually quite easy to jam local GPS receivers. A number of cases have been noted from around the world of explicit attempts to interfere with GPS to limit its usability. An article from 2012 describes different types of GPS jamming and countermeasures:

Types

- Unintentional interferences
  - Natural phenomena such as ionospheric disturbances and solar flares, ionospheric scintillations
  - Man-made phenomena such as inherent errors in satellites or signals and unwanted radio frequency transmissions due to television, microwave, communication traffic, or radar signals
- Intentional interferences
  - Jamming, meaconing, spoofing
- Meaconing involves interception and rebroadcast of navigation signals

Countermeasures

- System-level countermeasures
  - Multi-GNSS utilization
  - Signal authentication and encryption/decryption
- Antenna-based countermeasures
  - Adaptive beam-forming
  - Controlled radiation pattern antenna
- Receiver-based countermeasures
  - Time domain techniques
  - Frequency domain techniques
  - Transformed domain techniques
  - Switching frequencies in a multi-GNSS case
  - Integrating GNSS with inertial navigation systems
- Application level
  - Backup solutions: terrestrial backup
  - Comparison schemes to the predicted course of navigation (*Signal Barred*, from *Geospatial World*, November 2012, pp. 31–34)

While some interference types are clearly related to natural phenomena, a number can be directly connected to attempts to distort and limit the use of GPS. Clearly, there are cases in which the unexpecting GPS user needs to assess possible abuse or misuse of interference devices. People expecting either unintentional or intentional interferences should take note of countermeasures.

Wood's unveiling of the power of maps, we may be inclined to consider all maps as propaganda, with exceptions possibly for purely technical maps of infrastructure. However, propagandistic GI and maps probably influence our understanding in subtle ways, and broad application of skepticism such as Wood's may skip over those subtle influences. We are influenced in spite of ourselves. We should distinguish maps that empower us to engage and even manage aspects of our world from maps that subtly influence and bias our understanding and that therefore hinder our ability to act.

Summary

Cultural conventions guide the choices we make in our geographic and cartographic representation. The success of GI or a map is to no small degree contingent on the way it accounts for cultural conventions. Since a map or a set of GI emanate from a culture, their power is also a measure of that culture's engagement with the world. Understanding how culture influences GI and maps can aid in knowing how GI and maps are used and manipulated, and how biases are built in to maps. In Western-influenced countries a central issue is accuracy—questions of what and how GI and maps represent. A second central area of inquiry is how GI and maps came to be created. They are expensive to produce; therefore the economic and political impetus behind them should be examined carefully. Being aware of how cultures influence maps involves trying to understand how choices are made. How decision makers make trade-offs and perform cost–benefit analysis are applicable vocabulary. Awareness of such issues would also be relevant to ongoing debates about U.S. government surveillance.

Distortion is a common way to describe the inaccuracy of GI and maps. Propaganda is generally thought of as involving distortion of the truth, but distortion is a necessary part of most GI and cartography. Identifying possible sources of bias is how we might distinguish harmful from benign distortion.

REVIEW QUESTIONS

1. When is a map considered to be propaganda?

2. Why do maps and GI figure so significantly in our understanding of the world?

3. What are the three types of culture affecting GI and maps?

4. How can cultural issues and values influence maps and GI?

5. How have civilizations used maps?

6. What is an example of an indigenous form of mapping?

7. What is the main motive behind participatory GIS?

8. Why are people increasingly concerned about privacy protection and surveillance?

9. How can the privacy of an individual be impinged by surveillance technologies?

10. What is the difference between distortion and propaganda uses of GI and maps?

ANSWERS

1. When is a map considered to be propaganda?

   A map should be considered to be propaganda when the cartographic representation, geographic representation, or communication is malicious in intent. However, there is no bright line to distinguish propaganda biases from biases inherent in creating GI and maps.

2. Why do GI and maps figure so significantly in our understanding of the world?

   Maps are powerful visual ways to acquire information about places, things, and events we might never directly experience.

3. What are the three types of culture affecting GI and maps?

   The three types are national culture, indigenous culture, and disciplinary culture.

4. How can cultural issues and values influence maps and GI?

   They affect how people make meaning from GI and maps.

5. How have civilizations used maps?

   To the best knowledge of archaeologists, all civilizations have used maps to represent geographic things, events, and the relationships between them.

6. What is an example of an indigenous form of mapping?

   Stick charts used by Pacific South Sea islanders are one example.

7. What is the main motive behind participatory GIS?

   Participatory GIS seeks to support the needs of communities.

8. Why are people increasingly concerned about privacy protection and surveillance?

   Private companies and governments are collecting and combining more and more information about individuals.

9. How can the privacy of an individual be impinged by surveillance technologies?

High resolution remote sensing technologies could be used, for example, to detect the location of people at outdoor facilities.

10. What is the difference between distortion and propaganda uses of GI and maps?

Distortion reflects the unavoidable inaccuracy of geographic and cartographic representation to the actual situation. Propaganda distorts with malicious intent.

Chapter Readings

Cosgrove, D. (2001). *Apollo's eye: A cartographic genealogy of the earth in the Western imagination.* Baltimore, MD: Johns Hopkins University Press.

Craig, W. J., Harris, T. M., et al. (Eds.). (2002). *Community participation and geographic information systems.* London: Taylor & Francis.

Ghose, R., & Elwood, S. (2003). Public participation GIS and local political context: Propositions and research directions. *URISA Journal, 15*(APA II), 17–24.

Harley, J. B. (1989). Deconstructing the map. *Cartographica, 26*(2), 1–29.

Kosonen, K. (2000). *Kartta ja kansakunta: Suomalainen lehdistökartografia sortovuosien protesteista Suur-Suomen kuviin, 1899–1942* [The map and the nation: Finnish press cartography from the protests of the oppression years to the images of Greater Finland, 1899–1942]. Helsinki: Suomalaisen Kirjallisuuden Seura.

Monmonier, M. (1991). *How to lie with maps.* Chicago: University of Chicago Press.

Monmonier, M. (1993). *Mapping it out: Expository cartography for the humanities and social sciences.* Chicago: University of Chicago Press.

Monmonier, M. (1995). *Drawing the line: Tales of maps and cartocontroversy.* New York: Holt.

Pickles, J. (Ed.). (1995). *Ground truth: The social implications of geographic information systems* (Mappings: society/theory/space). New York: Guilford Press.

Pickles, J. (2004). *A history of spaces: Cartographic reason, mapping, and the geo-coded world.* New York: Routledge.

Rundstrom, R. (1995). GIS, indigenous peoples, and epistemological diversity. *Cartography and Geographic Information Systems, 22*(1), 45–57.

Sieber, R. E. (2000). Conforming (to) the opposition: The social construction of geographical information systems in social movements. *International Journal of Geographic Information Science, 14*(8), 775–793.

Thrower, N. J. W. (1972). *Maps and man: An examination of cartography in relation to culture and civilization.* Englewood Cliffs, NJ: Prentice Hall.

Toledo Maya Cultural Council and Toledo Alcaldes Association. (1997). *Maya atlas: The struggle to preserve Maya land in southern Belize.* Berkeley, CA: North Atlantic Books.

Turnbull, D. (1989). *Maps are territories: Science is an atlas.* Chicago: University of Chicago Press.

Turnbull, D. (1998). Mapping encounters and (en)countering maps: A critical examination of cartographic resistance. In *Knowledge and Society* (Vol. 11, pp. 15–43). London: JAI Press.

Wood, D. (1992). *The power of maps.* New York: Guilford Press.

Web Resources

⟳ The Integrated Approaches to Participatory Development (IAPAD) website offers a number of good resources and pointers to additional resources and discussions. See *www.iapad.org*.

⟳ The Aboriginal Mapping Network offers a number of good mapping resources. See *www.nativemaps.org*.

⟳ The British Library has an online exhibit that engages the role of propaganda maps. It is available online at *www.bl.uk/whatson/exhibitions/lieland/m0-0.html*.

⟳ An interesting collection of artistic engagements with many aspects of mapping culture is available online at *www.printeresting.org/2009/11/25/the-maps-as-art-exhibition-and-book*.

⟳ An example of using mapmaking with indigenous people is discussed and presented in a video at *http://google-latlong.blogspot.com/2012/06/surui-cultural-map.html*.

CHAPTER 13

Administration of Spaces

Defense, commerce, policing, and taxation are key government activities that require or benefit from clearly conceived and communicated GI and maps. Maps and GI are involved in almost every government activity.

After considering the impact of culture on GI and maps in Chapter 11, it follows that GI and maps for administrative activities and governmental policies would also be very important. Institutions regulate social relations, and doing so involves those social relations being represented in the form of GI and maps. Therefore, that geographic and cartographic representation can have an overt political character. At least the political role in administrative GI may be acknowledged, clear, stated, which we might contrast with the subterfuge of propaganda discussed in the previous chapter. A political objective is always present when social regulation is taking place. Nevertheless, maintaining political neutrality is sometimes a value in the culture of mapmaking. It may be a value even where there is no legitimate politically neutral point of view. Remaining neutral at that point puts one of the principal roles of geographic representation at risk, which is to help policymakers make better decisions.

A second political dimension of geographic representation in administration is that we "create" nature in the choices we make for representing natural objects. Deciding how to represent nature—and how to define it—is an important political activity that often involves the administration of spaces. From spotted owl protection to administrative regulations regarding the definition of protected plants and species, the geographic and cartographic representation of the natural world plays key roles in how people understand the world and act in it (see Figure 13.1).

Chapter 12 examined the different cultures and misuses of geographic representation and map representation in general and specifically considered

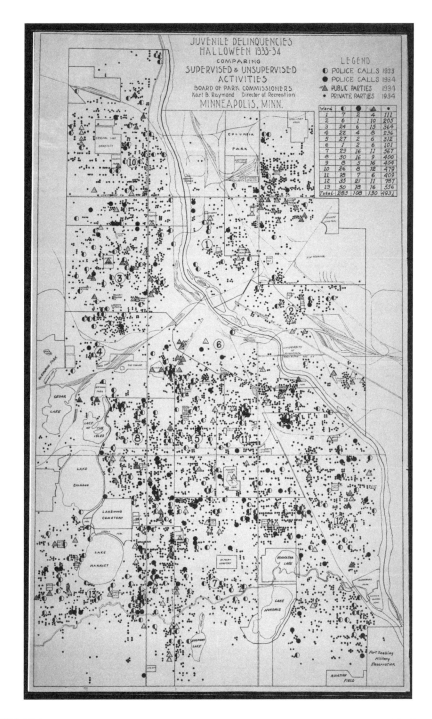

FIGURE 13.1. The map in this figure shows juvenile delinquencies on Halloween night for the period 1933–1934 comparing supervised and unsupervised activities in Minneapolis, Minnesota. This map helped police gain a geographical understanding of Halloween events in those two years and help plan activities during the year and for future Halloween policing. Some of the terminology is definitely out of date, yet suggests that cultural concepts of the time were very significant in categorizing events.

propaganda usages of GI and maps. This chapter in turn covers the empowering role and uses of GI and maps for administration. Clearly the topics of the two chapters have been and continue to be related. Building from the concept in the last chapter that culture always influences geographic and cartographic representation, this chapter begins with an overview of how GI and maps have become key references for the organization of so many social and economic activities. Administrative, especially governmental, GI and maps have created the world we know and continue to be key parts of how we understand and engage with that world. The chapter next considers the significance of land ownership registries—the cadastre—for capitalist economies and societies. The third section examines how administration activities have influenced the development of GI technologies. The closing section examines various sources of GI.

Administration of the World We Know

Through geographic and cartographic representations governments create spaces for economic, social, and cultural activities. The external boundaries and internal administrative divisions of the United States, Great Britain, Australia, Canada, Germany, Poland, India, and every other country have been created by many government administrations. Laws, regulations, and procedures rely on coordinate and locational systems.

Administrative activities in the 50 U.S. states provide both positive and negative examples of map use. From the settlement on the Eastern Seaboard and along the southern and western coasts by Europeans to relentless attempts to create advantageous election districts through gerrymandering, examples from the United States highlight the role geographic and cartographic representations play in our lives.

The history of the settlement of the colonies on the Eastern Seaboard shows that Europeans worked with maps to create the illusion that the land they were settling was relatively "unoccupied." As William Cronin describes the development of European settlements in *Changes in the Land*, setting boundaries for private land ownership and creating villages that could be readily mapped (in contrast to indigenous practices of sharing knowledge mainly through narratives and rituals) was a key part in establishing European administrations in North America.

The westward expansion of the United States points to the political importance of administrative activities that relied on geographic and cartographic representations. By failing to require surveyors of the lands brought under U.S. control to record existing indigenous habitation, early geographic representations rarely showed existing settlements and land use. This geographic representation suggested that vast areas of land were vacant and freely available.

The sharp angles of roads and agricultural land in these parts of the United States have always been interpreted from various points of view.

For some people, they indicate the rational underpinnings of the American economy and society. For others, the geometrical abstractions further remove traces of indigenous communities and much of the natural attraction of the land, subordinating it to calculative economics of gridded space.

Finally, the complexity of U.S. governmental administration results in short-term consequences being given priority over long-term ones. In the hierarchy of public administration across the 50 states, there are some 3,200 counties, further subdivided into 31,000 special departments (e.g., water, sewage, trash collection, fire protection). This makes local government in the United States very hard to navigate (see Figure 13.2). Resolving overlapping jurisdictional responsibilities has become an important obstacle in addressing many issues in the United States, most recently homeland security.

The Cadastre: Recording Land Ownership and More

Capitalist economies rely on the cadastre, which makes it one of the more important divisions of administrative spaces. The **cadastre**, a term that originated in Latin, refers to the registry of land ownership. In most places today

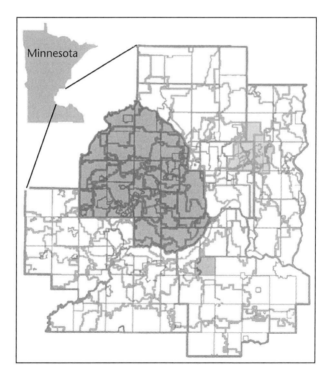

FIGURE 13.2. The boundaries of 293 public administration units in the Minnesota Twin Cities area. Coordinating the mapping activities of this many groups is very challenging but central to helping assure good services and government response.

the cadastre is fundamental to determining the rights and responsibilities of landowners and land users.

The cadastre consists of two parts. One part records the ownership of land in the form of a legal title, which also defines the land property. The second part describes the boundaries of the land property. The two parts are usually separately maintained—opening the door to substantial differences. This means that the land shown in a cadastral map need not match the land described by a cadastral title.

As a formal record of landownership, the cadastre is an important tool in both governmental administration and private enterprise. It serves to help governments identify who has to pay taxes, and what amounts, on their landownership, and also indicates the rights, responsibilities, and obligations of landowners and users. It is used by banks that provide mortgages, lawyers who draft contracts for land transactions, and companies that want to determine how land prices are changing in an area. In these ways, the cadastre is just one part of the broader socioeconomic uses of land called "land tenure." The Food and Agricultural Organization (FAO) of the United Nations defines *land tenure* as "the relationship, whether legally or customarily defined, among people, as individuals or groups, with respect to land." (See the FAO discussion "What Is Land Tenure?" in this chapter's Web Resources.) In contrast to the explicitly formal role of the cadastre, land tenure also includes individual agreements, such as when two neighbors agree to use parts of both properties as a shared driveway and place to park their cars, or when an older person in the family allows younger family members to use the property without charging them money or selling the property to them. It also involves the more complex relationships between people that involve land.

The cadastre records landownership and through its formal significance accomplishes or certainly helps to accomplish much more, but it is only part of the relationships people have to the land they live on or use. The cadastre is a key part of any capitalist economy, but its relevance depends to a great extent on its tangibility for day-to-day relations. Access to GI is important. Opening data access online can have complex consequences. Many of those who would benefit probably won't, because they lack skills, knowledge, and resources.

In the United States, most landownership is recorded using the Public Land System (PLS) (see Chapter 5 for more specific details). This cadastral system guarantees a relatively simple system of land title and boundaries, but errors and complex situations nonetheless still arise. Many states in the United States require title insurance on all land transactions to cover the risk that some other valid claim might be made related to an area of land. In this system any maps or plans showing the boundaries to the title are incidental (see Figure 13.3).

On the East Coast and in a few other areas of the United States (Texas, Louisiana, and parts of California and New Mexico) the metes-and-bounds system dominates. This system is more involved for surveyors and can be very expensive if differences between neighboring property owners must be

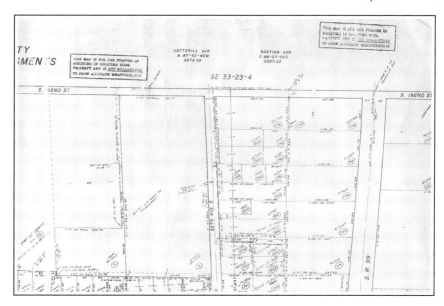

FIGURE 13.3. Portion of an assessor's map. Such maps are designed to facilitate communication and coordination among professionals working with the cadastre. Note the disclaimer on the top of the map. It offers guidance to try to avoid misuse of the map.

resolved in court. The Torrens system, developed by Sir Robert Torrens in Australia in 1857, provides for a clear registration of property and creation of a title. The title does not transfer to later owners; it must be reregistered with the court. In the Torrens system maps and plans are integral to the title.

Administration Impacts on GI Technologies

Government administrations have had and continue to have significant impact on the development of GI and maps, particularly through their role in requesting and financing the development of new technologies. Obviously, the complexity of cadastres means attempts should be made to improve administrative technology (see Figure 13.4). Many of these efforts have focused on internal improvements and are scarcely noticed by the general public, except when permits and taxes become more transparent. These internal improvements have focused on two areas: improving administrative activities through the use of GI technologies and improving administrative coordination between different agencies. Chapter 15 examines some of the successes in improving administrative activities. In the current chapter, we focus on how government coordination has been improving because of administrative-led developments of GI technologies.

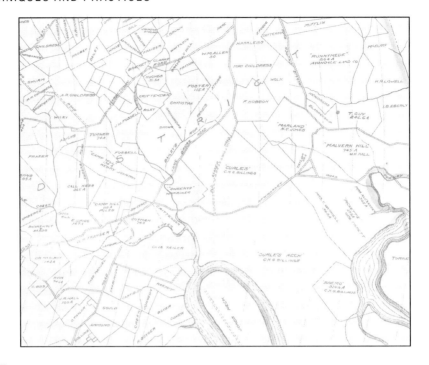

FIGURE 13.4. Cadastral map, Henrico County, Virginia, 1916. The cadastre in many countries follows strict legal mapping guidelines.

Multipurpose Land Information System

Because improving administrative activities can be greatly aided by improving data sharing, many GIS-related administrative developments have focused on improving data sharing, not only through technological standards, but by establishing a common reference base for administrations. The Multipurpose Land Information System (**MPLIS**) is perhaps the best known and most significant of all the concepts for improving administrative data sharing.

MPLIS has its roots in cartographers' attempts to develop multipurpose cartography in the 1950s. *Multipurpose cartography* relied on the preparation of different thematic layers using a common coordinate system that facilitated the manual overlay of the layers in the preparation of the printed map. The thematic layers corresponded to administrative activities—for example, U.S. states as a base layer combined with another layer of highways or a layer showing railways to produce maps showing different parts of the transportation infrastructure.

MPLIS starts with this multiple-use concept and simultaneously extends it and focuses it on local governments, counties, or municipalities. The extension of the multipurpose cartography concept focuses on making the resulting combined information the catalyst for administrative coordination. This is possible because the GI technology developed since the 1960s made it very easy to combine data as long as a common coordinate system was used. To make maps for presentations and publications a cartographer was often still called for, but maps for administrative analysis or straightforward informational purposes could now be produced by people working in administrative offices instead of specialized cartographers.

For the MPLIS to work, a great deal of coordination and data sharing is required. Administrative resistance is most often the greatest stumbling block for data sharing. Conflicts between different mandates and different information needs often impair the development of the MPLIS in local governments. The many success stories around the world point to the great significance of this administrative concept for the improved efficacy of administrative cartography and GI.

Spatial Data Infrastructures

In some senses spatial data infrastructures (**SDI**) are extensions of the core MPLIS concepts (coordination and data sharing using GI stored in a common coordinate system) to other levels of government. The first, and still most significant, development of the SDI occurred in the United States through the creation of the National Spatial Data Infrastructure (NSDI).

Extending the core MPLIS concepts of coordination and sharing to larger geographic regions, the basic principle of the SDI is that freer access to administrative GI contributes to good governance and a better society, government, industry, and people. The MPLIS provided evidence for the potential; the SDI grapples with the issues of developing the capabilities of networked data sharing and administrative coordination that use computer networks to transfer data between different sites.

Because the NSDI was conceived while the Internet was being developed, its underlying concepts work without a network; its implementations explicitly rely on the Internet. The initial goals for the NSDI focused on civilian federal agencies and emphasized the importance of these agencies sharing GI as a means to reduce the expenditures for the federal government. Although the NSDI was conceived of as guidelines for the development of infrastructures that reached from federal agencies to local administrations, it was only sporadically implemented as intended because the NSDI lacked legislation that broadly supported the involvement of local governments or required the participation of state and local administrations. On this note, it is important to recognize the inspiration and conceptual guidance the NSDI has provided administrations in the United States and other countries. While perhaps not realized as intended, the NSDI has been an important driving force for improving interoperability.

The NSDI in the United States consists of four components: framework, vertical and horizontal dimensions, metadata, and the availability of free or low-cost data. The framework defines seven data layers needed for nearly every government activity. This data should provide nationwide coverage. The framework has both vertical and horizontal dimensions. *Vertical* refers to data exchange and sharing between different levels of government;

horizontal refers to the exchange and sharing between administration units at the same level of government. *Metadata,* or data about data, is crucial to helping people and computer searches in quickly finding GI and evaluating the suitability of the GI for their purposes. Finally, the NSDI includes the concept that framework data should be *freely available.* Other data can be freely available, or may be only available after paying a fee or registering with the data provider. This theoretically allows for the greatest reduction of data collection and maintenance costs by providing support for as many different GI applications as possible.

Other SDI were developed in the United States and around the world after the U.S. NSDI was proposed. They directly incorporated networking capabilities in their design and implementation. These SDI also are largely developed to support primarily intragovernmental data exchange and sharing. Public access is a component but often quite limited. Most common is to allow access to Internet services that facilitate the creation of maps (see Chapter 14). Some SDI are less integrated in the NSDI. They offer services to examine descriptive data about available data sets (metadata) or, after registration, to log on to a protected website and download data. The European Community INSPIRE directive is one of the most successful and advanced approaches drawing on SDI concepts active in the world today.

Free, simple data sharing is the central concept for most U.S. federal government SDI, but has been greatly constrained even here and in every instance by financial and participation issues. Data sharing is made difficult by problems allocating and distributing costs, but institutional and political issues related to participation play a sizable role as well. While there is no conclusive evidence that charging for GI provides enough revenue to recover the costs and maintenance of the GI, or even the costs of managing the system for regulating use and charging, most administrations around the world continue to charge, even in cases of use by other administrative agencies. People voice concerns about paying for the data necessary to assure public safety as well as civil and environmental protection and administrative budgetary concerns, but the developments of SDI have been greatly impaired by cost recovery. SDI remains one of the key concepts guiding developments of GI. Where and when successful, it is possible to witness the profound impacts of spatially enabling society. How individual regions and countries address the challenges is an important question for the future of GI and cartography.

In considering the past, present, and future of SDI, it is important to note the important roles of standards and specifications. Both are terms with specific meanings in a clear context, but generally tend to be more ambiguous. Both provide guidelines for the development and implementation of SDI and GIS. Standards tend to have more legal or governmental backing, although this is by no means exclusive. Specifications are also guidelines, but reflect commonly used approaches, or industry-accepted best-practice formulations of specific requirements and/or guidelines. In general use the terms are used interchangeably with great frequency, which can lead to some confusion. In every case, standards are definitely good; the adaptation of standards for specific needs and institutions necessitates later the resolution of different standards' adaptations. In the United States, the Federal Geographic Data Committee has played an important role in developing federal government standards. State, tribal, regional, and local government standards and specifications continuously play a role in improving the administration of GI through standards. They are frequently connected to national and international activities.

Digital Libraries

Work on digital libraries addressed a very important source of information for society: public libraries remain great repositories of information and offer support for knowledge economy activities that find little support elsewhere. Compared to traditional map libraries, digital map libraries offer some capabilities that are of great benefit to many users of GI and maps. The key difference is that digital map libraries combine both paper maps and GI.

This can occur through the scanning of existing paper maps (when possible) and their storage in a digital format. Even with referencing to a coordinate system, scanned paper maps are still cumbersome in comparison to GI. However, when copyright laws allow it, scanning provides a straightforward way to collect information. Digital libraries of GI also support novel methods for accessing GI that reflect the complexity and the variability of how people work with and use GI and maps. Digital libraries can distribute the physical storage of GI to different sites. The possibility also exists of accessing information from multiple sites for digital libraries. A person accessing a university digital library of GI and maps may transparently access GI and maps from other university libraries.

Digital Earth

Proposed in the late 1990s, Digital Earth permits a person anywhere in the world to access GI for any place on the earth, at variable scales and resolutions, via the Internet. The data is visualized simply on a virtual globe. Digital Earth has not been used much by governmental agencies working with GI, but it has certainly been important for agencies involved in global issues and for providing three-dimensional visualizations of many areas of the earth. Google Earth, Bing! Maps, World Wind, and many others draw on concepts from Digital Earth, so the unparalleled success of those online mapping services, in many ways, has been the success of Digital Earth. The way that many people around the world approach GI and mapping has been substantially altered.

Digital Earth also exists as a reference model specification for the global georeferencing of GI. This mainly supports global environmental research, but, clearly, it still can be a valuable reference framework for other uses.

Research Support

The projects listed in the preceding sections have, in general, been developed with a relatively small amount of financial support. Despite the small amounts of seed money, the MPLIS, NSDI, and Digital Earth have been used for a wide range of purposes. Governments also have specifically funded improving techniques for collecting and maintaining data, assessing and improving administrative services, and developing new technologies and approaches to make better use of GI and maps. This support has been crucial at various points in the development of the over U.S. $200 billion/year GI and cartography industry worldwide.

Government Sources of GI

Most available GI comes from government agencies. Sometimes it is not easy to find GI for a specific region. This section only attempts to overview key sources. For specific GI needs, you should first contact a nearby map library or government agency that shares GI.

North America

The United States seems to stand out globally because of the widespread ease of obtaining data, but that is actually certain only for most data held by federal civilian agencies. Public domain data collected by civilian agencies of the U.S. government with general funds are considered to belong to U.S. taxpayers. All states have their own regulations, generally called "open records laws," which describe which data can be made available, under what restrictions, and at what costs.

The U.S. National Atlas (*http://nationalatlas.gov*) is a good starting point. It supports the online creation of maps as well as data downloading. Data about roads, address ranges, and census geography is available at *www.census.gov/geo/www/tiger*; however, this data requires some processing before it can be viewed and no online browser is available at this site. Some remote sensing imagery can be found at *http://rsd.gsfc.nasa.gov/rsd/RemoteSensing.html*. Additional land remote sensing data and individual images are available at *http://eros.usgs.gov/archive/nslrsda/*.

The U.S. states maintain individual websites for getting information and accessing GI and making online maps. There are far too many to list here. You can go to the geospatial portal of *geo.data.gov* for a list of data and access information (*http://geo.data.gov/geoportal/catalog/main/home.page*). Local governments also may maintain online access possibilities. You should be able to find these out by going to the home page of the local government (county or municipality) in question.

In Canada, the Canadian National Atlas (*http://atlas.gc.ca/site/english/index.html*) is an excellent resource, as are increasingly the provincial, county, and municipal governments. Geogratis and GeoBase provide access to large amounts of data in Canada following the Government of Canada's commitment to open government. GI for Mexico is far scarcer.

Europe

The GI available in Europe tends to be only available for government agencies or only after purchasing. A good starting point for European environmental data is the UNEP-GRID office in Geneva (*http://ede.grid.unep.ch*). Another good source is the European Environment Agency's data service (*www.eea.europa.eu/data-and-maps*). At this site you can browse metadata and if you sign an agreement you can also download data.

The sources for GI from various countries and regions vary greatly. For instance, Danish GI is available online at *www.gst.dk/English/The+Spatial+Data+Infrastructure*, but it appears that beyond making online maps that data must be purchased. The U.K. Ordnance Survey has a great deal of GI, which can be accessed for commercial use after applying and usually paying a fee (*www.ordnancesurvey.co.uk/oswebsite*). Educational uses in the United Kingdom are often covered through specific low-cost licenses. Poland makes some data available over the Internet for viewing only (*http://maps.geoportal.gov.pl*). Other sources for European data can be found at the European Umbrella Organization for Geographic Information (EUROGI) website (*www.eurogi.org*). The GeoNetwork (*http://geonetwork-opensource.org*) lists links to many national and regional administrative sites that provide GI access in Europe. Finally, OpenStreetMap (OSM) data is available for much of Europe, as well as the rest of the world.

Other Parts of the World

Beyond OSM GI, some GI for New Zealand is available (*http://geodata.govt.nz*). The Australian Spatial Data Directory (ASDD) provides search tools to search for geospatial data set descriptions throughout Australia (*http://asdd.ga.gov.au/asdd/*). There are many sources of data for Eastern and Southern Asia. An accessible source of GI for China is provided by the University of Michigan China Data Center (*http://chinadataonline.org/cge/*). A list of administrative sources of Brazilian data (*www.inde.gov.br*) provides a very large set of potential GI sources. INPE, for example, provides a portal to access their information (*www.inpe.br/acessoainformacao/*). Other areas of the world have a great deal of GI, but most readily available data is part of global or regional data sets. The Group on Earth Observations provides a portal to search for global data (*http://geoportal.org/web/guest/geo_home*), as does the Food and Agricultural Organization's GeoNetwork (*www.fao.org/geonetwork/srv/en/main.home*). Of course, this is just an arbitrary selection intended to provide a basic overview.

Summary

Mapping and GI empower the creator and user. Maps are used in almost every government activity because of the power of what maps show. Governments use geographic and cartographic representation to create spaces. The cadastre is a prime example of how administrations in Western civilization have created spaces. No matter how much effort the administration puts into a cadastre, it must be accepted that its boundaries are related to actual boundaries to be relevant. The MPLIS was one of the key post–World War II information technology concepts for administering local government activities. The SDI develops these concepts to involve multiple governmental units and benefit from computer networking. In many cases, the adoption of data sharing and coordination policies has led to much broader spatial enabling of society. Other concepts that have advanced the ways of administering spaces are digital libraries and Digital Earth. The results of these concepts can be seen in the increasing access to GI around the world and improved representations and communication.

REVIEW QUESTIONS

1. Why is there such a large administrative interest in GI and maps?

2. What are the two ways of recording the boundaries of land ownership?

3. What does the abbreviation PLS stand for?

4. What is the basic principle behind the SDI?

5. How do coordinate and locational systems become key references?

6. What is the definition of *cadastre*?

7. What makes data sharing so difficult?

8. What does "public domain GI" mean?

9. Explain the key principles of the MPLIS.

10 To which parts of government in the United States does the public domain principle apply?

ANSWERS

1. Why is there such a large administrative interest in GI and maps?

   Defense, commerce, and taxation are key government activities that require or benefit from clearly conceived and communicated GI and maps.

2. What are the two ways of recording the boundaries of land ownership?

   The most common approaches to recording the boundaries of land ownership are metes-and-bounds surveys and systematic surveys (e.g., PLS).

3. What does the abbreviation PLS stand for?

   PLS stands for Public Land Survey. "System" is often appended to create the abbreviation PLSS.

4. What is the basic principle behind the SDI?

   The basic principle of the SDI is that freer access to administrative GI benefits good governance and society, government, industry, and people.

5. How do coordinate and locational systems become key references?

   By becoming part of administrative laws, regulations, and procedures, coordinate and locational systems become part of social interactions and help structure the human-influenced geography of an area.

6. What is the definition of *cadastre*?

   Cadastre is a Latin term that refers to the registry of land ownership. In most places today that understanding is extended to include the role of ensuring the rights and responsibilities of landowners and land users.

7. What makes data sharing so difficult?

   Data sharing is directly made difficult by problems allocating and distributing costs, but institutional and political issues play a sizable role as well.

8. What does "public domain GI" mean?

   Data collected by civilian agencies of the U.S. federal government with general funds that are considered to belong to U.S. taxpayers are generally available for no or little charge.

9. Explain the key principles of the MPLIS.

   The Multipurpose Land Information System (MPLIS) shares the GI from county or municipal agencies by using a common coordinate system. Some systems additionally use the parcel, or cadastral, "layer" as the base layer to align other layers.

10. To which parts of government in the United States does the public domain principle apply?

    The public domain principle in the United States only applies to civilian federal governmental agencies. State and local governments follow the open records laws of their state.

Chapter Readings

Cho, G. (2005). *Geographic information science: Mastering the legal issues*. New York: Wiley.

Chrisman, N. R., & Niemann, B. J. (1985). *Alternative routes to a multipurpose cadastre: Merging institutional and technical reasoning*. Paper presented at AutoCarto 7, Washington, DC.

Cronon, W. (1983). *Changes in the land: Indians, colonists, and the ecology of New England*. New York: Hill & Wang.

National Research Council. (1980). *Need for a multipurpose cadastre*. Washington, DC: National Academy Press.

National Research Council. (1990). *Spatial data needs: The future of the national mapping program*. Washington, DC: National Academy Press.

National Research Council. (1994). *Promoting the National Spatial Data Infrastructure through partnerships*. Washington, DC: National Academy Press.

Sherman, J. C., & Tobler, R. W. (1957). Multiple use concept in cartography. *Professional Geographer, 9*(5), 5–7.

Tulloch, D. L., Epstein, E. F., & Niemann, B. J., Jr. (1996). *A model of multipurpose land information systems development in communities: Forces, factors, stages, indicators, and benefits* (GIS/LIS '96). Denver, CO: ASPRS/AAG/URISA/AM-FM.

Web Resources

↻ For an FAO discussion of land tenure, see *ftp://ftp.fao.org/docrep/fao/005/y4307E/y4307E00.pdf*.

↻ For an example of unexpected consequences from digitizing cadastral material and opening access, see *http://crookedtimber.org/2012/06/25/seeing-like-a-geek*.

↻ For an adaptation of the MPLIS concepts for the U.S. Bureau of Land Management, see *www.blm.gov/nils*.

↻ A starting point for information about the U.S. NSDI is available online at *www.fgdc.gov/nsdi/nsdi.html*.

↻ Information about individual U.S. state open records and Freedom of Information Laws is available online at *www.nfoic.org/state-freedom-of-information-laws*.

↻ The European Commission's INSPIRE guides the creation of SDIs for environmental programs. See *http://inspire.jrc.ec.europa.eu*.

↻ The Global Spatial Data Infrastructure (GSDI) is an organization promoting the development of SDIs around the world. See *www.gsdi.org*.

⟳ See information about Digital Earth concepts realized through the Planetary Skin Institute at *www.planetaryskin.org*.

⟳ A conceptual heir to the Digital Earth concept is Google Earth, available online at *http://earth.google.com*.

EXERCISE

Why Should GI Be Expensive?

GI can be hard to obtain and expensive to produce. Why is that? Consider both user and provider perspectives.

Should GI be made as cheap as possible or should it be sold for enough money to cover the costs of collection and maintenance?

CHAPTER 14

Online Mapping
and Geocoded Worlds

The Internet has had a huge impact on communication technologies, and GIS are no exception. This chapter engages with the mushrooming possibilities of online mapping and geocoded data. While many GIS functions have been implemented using the Internet, fundamental changes that come with easy access to online information lead to new possibilities for mapping. This chapter focuses on these issues and turns to the new significance of location in the information age that comes with geocoding. It also considers the collection and use of location data collected by mobile devices and software, often without the owner's explicit knowledge in terms of privacy. Finally, this chapter considers what Big Data is and means for future users of GI.

Online GIS

First, in the 1970s and 1980s, there was GIS, and people used GIS by going to a computer with the GIS software already installed, or they worked with specialists trained in the GIS software. This was better than going to a map library to look for a map and was already much faster. Then came the Internet, particularly from the mid-1990s on. Innovators around the world were quick to realize that one computer with installed GIS software could be used to produce maps by people at other computers in other offices, in other buildings, in other parts of the city or state, even in other parts of the world. Online GIS had started.

The development of online GIS followed on the heels of the development of the Internet. The first versions of the Internet, ARPANET, supported by the National Science Foundation and other U.S. government agencies, allowed a client computer to send a message to a server that completed the request and returned the result to the client. Analogous to a restaurant, remote procedure calls (RPC), as they were called, made it possible for one server to help many clients access the services they wanted. The first concepts for online GIS designed ways to handle requests for maps. For example, a programmer could design a service to produce neighborhood maps. A person at a client computer could request a map showing the Johnson neighborhood census data. The server would take the request, determine the extents of the Johnson neighborhood, produce the map using predefined specifications, export the map to a format that the client could use, often the still ubiquitous PDF format, and send the file to the client. The client could open the map. While this approach wasn't so fast as to be called "interactive," it was far, far faster than previous arrangements. And it was hugely successful in making it possible for many people to get maps made to fit their needs and help answer their questions in a much faster manner than people just a few years earlier could have imagined.

The boom in online GIS of course reflected the mushrooming of the Internet and nascent World Wide Web in the late 1990s, and led to a virtual explosion of services not only to expand the online creation of maps, but also to integrate data collection, analysis, and communication, and create new types of operations that allowed for some measure of coordination between offices and remote sites. Advances in technology, particularly the enhancements made to improve data transfers with mobile devices, have made it possible to provide handheld devices with GIS functionality.

Also, web GIS has grown immensely. From creating maps, converting them to PDF files, and sending them to a client, web GIS mapping now involves an uncountable number of map server-type software systems for developing spatially enabled Internet applications. Some of these systems are open source, meaning the software is distributed for free and is free to access, and some are proprietary, requiring purchase of the software. These systems support the creation, management, and use of applications online. Some of the online resources listed in the Web Resources section at the end of this chapter are a good starting point to enter the rich and complex world of online GIS. They should help you find a number of online GIS resources in your community, county, state, or national government offices.

Online Mapping

Many groups have developed online GIS, but it seems likely that online mapping activities dwarf it considerably. We should distinguish online mapping from online GIS in two ways. First, online mapping software originates from earth viewer software developed to provide satellite and remote-sensing, image-based dynamic interactions for a user with GI. Second, additional data and choices for symbology of the online map can be made directly by users in a variety of ways. The second point also refers to mash-ups, a web-based application that takes data or functionality from other online sources to create new data or functionality. While online GIS applications can have similar capabilities, the abilities have to be designed and created by individuals or groups with the proper, and often limited, access to the software and servers. Online mapping, in distinction, allows for broader uses and open-ended development in its very design (see Figure 14.1). Many established groups prefer online GIS because of the greater control it provides. Online mapping provides possibilities for users that can be compared to an encyclopedia of geography, or a geographic Wikipedia, allowing people to access existing GI and create their own resources, data, maps, and even functionality (see Figure 14.2).

The one billion downloads of Google Earth, which occurred in October 2011, is a number whose sheer size points to the widespread acceptance of the online mapping approach and the persistent desire to enable individual decisions about how and what to map. Other online mapping packages may have smaller numbers of downloads, yet they too have had significant impacts in particular application domains or geographical areas. Taken together, all online mapping software has become the central way many people now first see a map in their lives, or regularly use and now make maps to answer questions they have about the world, their community, their country. The uses of online mapping are, in this sense, only limited by people's imagination.

An important problem that arises with the boom of online mapping, as with many things on the Internet, is finding data, maps, and functionality. While the facts for this quick math are not perfect, even if just 10% of

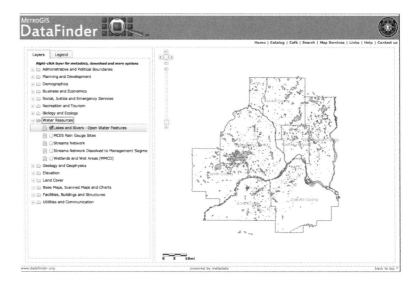

FIGURE 14.1. The Twin Cities Metropolitan area supports many GIS-related services and operations, especially coordination and data access. DataFinder is an example of a portal to GIS data used by professionals, government staff, and others to gain access to data for the area.

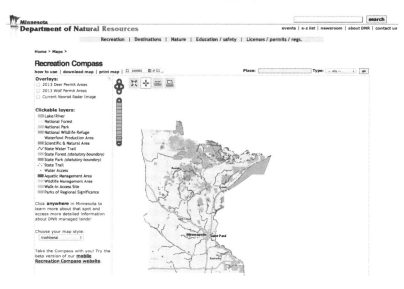

FIGURE 14.2. The Minnesota Department of Natural Resources (DNR) provides an online mapping tool to allow people interested in Minnesota recreation and outdoor activities information to help in their planning and organization of trips.

the people who downloaded Google Earth made one data set, it still means 10 million data sets would have been created. How do you find what you're interested in with that many data sets? The most common answer, which is also the means to find online GIS data, is widely known as the *geoportal*. Geoportals organize information from diverse sources in a systematic way (see Figure 14.3). Usually they are created and maintained by government agencies, nonprofit groups, or commercial companies to help staff, the public, or customers find GI to match their interests. For example, a county can have a geoportal to list and describe the various kinds of GI available from different county agencies; a company that has many sites could use a geoportal to help suppliers get specific GI about each site. A geoportal can function very

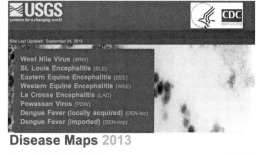

FIGURE 14.3. Government agencies provide important mapping resources. The United States Geological Survey (USGS) provides online mapping support to make maps of widespread diseases, among many other types of online mapping services and support.

simply as a list of related sites and sources for GI, or add functions to support queries by geographic region, type of data, keywords, people, etc. Usually simple geoportal design is most effective, but the simplicity cannot limit the usability for most of the purposes people will access this GI for. A key group in supporting the development of specifications and discussions of guidelines for online mapping is the Open Geospatial Consortium (OGC). Standards are crucial to enhancing collaborative uses of online mapping, but can become a dilemma between promoting cooperation and locking in arrangements that preclude or limit access to others.

Geocoded Worlds

Online GIS and online mapping are crucial technologies for making information connected to a location available online. While many people have prophesized the death of distance or a flat world that comes through the widespread and increasingly ubiquitous availability of online resources in the information age, the opposite may be closer to the truth, although far more subtle. Indeed, location is becoming ever more important. Location, of things and events, is of paramount significance in finding relationships and combining data from multiple sources to gain insights into things, events, and the places where they are located or happen.

For people to have benefit from location data, they need to have data with location information. Geocoding, from a technical point of view, is the process of determining location in a geographic reference system from other data with a different spatial reference. A common example is house addresses. We regularly use addresses to locate stores, places, and residences by themselves. They are rarely perfect, and without familiarity can even be very confusing, but they are conventional arrangements that are extremely flexible and useful for most day-to-day uses. This flexibility that has provided so many advantages for hundreds of years, however, leads to a number of problems when people begin to rely on computers to locate things and events. With some effort, it is possible to reliably transfer data from street addresses, postal codes, census blocks, watersheds, and the like through the geocoding process to systematic location information of a geographic reference system, frequently latitude and longitude coordinates (see Chapter 6). Once the information has been made, in this sense, geographic, it can be combined with other location data to support any number of mapping and analysis activities (see Figure 14.4).

The frequently used term neogeography, shortened to neogeo, reflects the increasing role of geocoded information outside the traditional domains of cartography, GIS, and GIScience. (For example, countless Internet applications and device applications, such as restaurant finder apps, use *location* as the key dimension for relating things and events to one another.) While initially used to point to alternative cultures and novel approaches to geography taking place in the arts and outside the established field of geography,

FIGURE 14.4. Infoamazonia is a nonprofit group that provides mapping services and information to help people in Amazonia and other parts of the world understand environmental and economic changes in Amazonia. The stories it provides help people gain better understandings of these changes through the presentation of specific issues.

neogeo began to take on a more technical meaning. Still, though, it did avoid making reference to specific technical concepts such as scale and spatial analysis that were at the core of GIS and GIScience. Over time it has become a means to distinguish crowd-sourced GI from authoritative GI, and increasingly it refers to GI for consumer-oriented applications. These applications often rely on geospatial information processing outside of a GIS system and are used in the context of open-source development. While the resulting visualizations can be evocative, the challenges of connecting GI analysis to neogeography are considerable and often require specialized programming capabilities (see Figure 14.5).

Location Privacy

The developments of online GIS, mapping, and the geocoding of the world have gone hand-in-hand with the largely unregulated collection of data from the over 6 billion cell phones in the world and other sensor devices. For example, it is increasingly likely that during shopping at larger stores the location of your activities was recorded and used with data from other shoppers to assess what was most attractive and kept peoples' interest the longest. While this data is usually anonymized and no personal data is collected, it is theoretically possible using other data, for example, credit card point-of-sale transaction data, to establish with great likelihood who was in the store

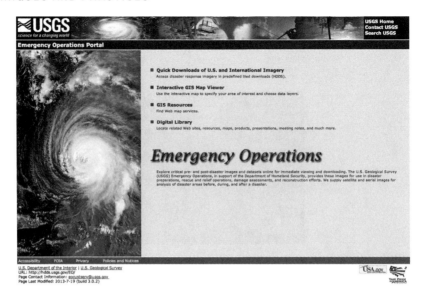

FIGURE 14.5. Many online mapping resources provide access to data and online maps in different forms. The USGS Emergency Operations website provides imagery downloads, an interactive map viewer, GIS resources, and a digital library with additional information.

around a particular time. Clearly, even this theoretical potential would give many people reason to be concerned about the increases in data collection and the potential loss of their privacy, specifically privacy in the geocoded world, or location privacy.

In many areas of the world, the increased surveillance through stores, public transportation, companies, utilities, neighbors, and the like has become much more commonplace and accepted (see Figure 14.6). In many settings people have come to even find a degree of safety in the surveillance, but some people have begun to raise questions about the amount of data collected by government and private agencies and what is done with this data. As with information collected by product registration forms, applications, or using the phone or mail, which can be aggregated using common identifiers—for example, in the United States the Social Security number or house address—GI can also be combined to create more detailed analyses.

The "locational data" of everyday life—when and where people go shopping, when they walk the dog, where they go for a run, and when and where they meet friends—people consider this to be private information. New technologies make it easier to track movement and collect this kind of individual-level information. Public safety concerns may make the collection of this information necessary for particular situations, but the widespread collection of individual data and its uncontrolled storage and use concerns many people. With new technologies, many discussions around the world are unfolding about how to adequately protect privacy, yet at the same time ensure that adequate information is collected for law enforcement and defense purposes. The legal and institutional decisions hinge very much on cultural conceptions of geographic and cartographic representation of individuals.

New surveillance technologies are pervasive and subtle. From drone aircraft that can stay aloft for extended periods to small computer chips the size of a grain of rice, the development of surveillance technologies is making it easier to record information about individual movements without the knowledge of the person under surveillance. Combined with GPS technology, it is becoming possible for many people to assemble data into a study of how individuals spend their time. The Apple LocationGate scandal in 2011 illustrated this problem for millions of people.

The first of these technologies in terms of relevance for recording individual movements and situations is satellite remote sensing. High-resolution remote sensing technologies offer resolutions of 61 cm, allowing for the detection of the presence of individuals in clear situations—for example, while standing on a concrete parking lot. Airplane collection of aerial photography can readily record 6-inch resolution data. Satellite remote sensing can be complemented and enhanced by ground-based remote sensing technologies. Britain has installed over 2.5 million surveillance cameras, offering many possibilities to record the movements of an individual. These technologies, which offer direct observations, can be easily combined with data from technologies that routinely collect information about the location and movements of people. Lastly, users of cell phones can be readily localized

FIGURE 14.6. While many government agencies provide online data and mapping, the costs for collecting and preparing data have become low enough to allow groups of people to collect their own data (e.g., Open Street Map, *osm.org*) and increasingly even more detailed data. This picture shows part of a car equipped with sensors to record continuous georeferenced imagery from the roof of a car.

and phone records can be used to ascertain the general movements of a cell phone user, whether they are making calls or not. Other means arise from the increased use of electronic payment systems.

These data from these technologies can be combined to create detailed profiles of individuals. This has been done for a number of surveys. The use for criminal investigation has been popularized in movies and TV shows. The mundane use of personal data to aggregate information for marketing and other purposes has received some coverage as the result of the increasing number of identity theft cases, but its use by telemarketers and direct mailers has lacked a clear response.

The complexity of an individual's movements and the limitations of the technologies lead to constraints. GPS receivers will not work inside buildings, and most high-accuracy remote sensing technologies will not record details about objects in the shadows of buildings or obscured by natural features (e.g., in ravines or under dense tree coverage). In other words, some unsurveyed areas remain, but as the technologies develop and people come to rely more on them for a sense of security, the cultural sense of what is appropriate to collect, record, and reuse about an individual's things and events will be changing.

Spatial Big Data

Given how often Big Data has been discussed in recent years, it is surprising that it is seldom discussed in the context of GIS. However, the prospect of

Privacy: An Evolving Concept

Changes in technology have always led to changes in the concept of privacy. The printing press, the camera, the phone, and the mobile phone have all led to changes. Technologies allow for changes in how people and the government can make observations and require modification of laws and regulations to protect social and cultural conventions.

Presently, the boom in mobile technologies, on the heels of Internet technologies' widespread impact, is leading to a number of privacy concerns around the world. The related changes have led to different ways to think about privacy. The philosopher Helen Nissenbaum examines these issues and comes to the conclusion that we should think about privacy in context. What she means is that our sense of privacy is connected to a particular context. Some of what defines a context is physical, some social, some cultural, some individual, and so on. We have different expectations of privacy at home and at a sports event. With all this variability, it seems that location is an important characteristic of privacy.

In the absence of any universal right to privacy in almost every jurisdiction, evolving concepts of privacy will continue to attract great attention, require the input of specialists working on a range of issues, and depend on the same specialists to help communities understand the evolving sense of privacy in the information age.

using Big Data in GIS and spatial analysis holds great potential and some pitfalls. It is valid to ask whether GIS would have to change in a fundamental way to handle Big Data. One might think that GIS would not change, because there are millions of users, GIS' handling of a vast range of data is versatile, and GIS have extremely powerful spatial analysis capabilities. But considering how much data being collected today has a locational component, that would be a short-sighted assessment. While no full Big Data GIS is yet available, we already are starting to see changes in the development of techniques that make it possible to import and export data between GIS and Big Data databases. This development, even by itself, might be a stepping-stone toward the creation of Big Data GIS. This is often known as location analysis and is widely used with data coming from mobile device users.

This section sets out to simply introduce interested readers to what may transpire in the efforts to connect Big Data with GI, as well as pointing out some of the better-known and lesser-known challenges and concerns.

Before getting to these, let us first consider what Big Data is and what possibilities it already has shown for GI analysis. We need to review the central concepts of Big Data in their context and relate them to GI.

What Is Big Data?

The obvious point to start from is to define Big Data. There is, however, no single definition, indicator, or scale that gives total clarity. Opinions vary. The most commonly used description is that it involves data sets so large they are beyond the capabilities of widely used software to process, manage, and even capture in an acceptable amount of time. What is beyond those capabilities may be a matter of perception, and computational power increases rapidly, anyway. A helpful way to understand the size aspect is in terms of volume, variety, and velocity—the three V's. Considering the big picture, the reality is that much data collected today is already so staggeringly, even inconceivably, large that it already is Big Data. Processing it already requires a different approach. Considering the growth of data on a global scale, some researchers calculated that in 1 day people create 2.5 quintillion (25,000,000,000,000,000,000) bytes of data. That data put on DVDs would produce a tower of DVDs reaching from the earth to the moon and then another tower reaching all the way back. That's definitely Big Data! Now it is true that other estimates of the amounts of data produced daily vary, but they are all of magnitudes beyond most imaginations. Even individual measures of specific kinds and uses of data are staggering. For example (based on numbers from 2014), if you use Twitter, each tweet is one of 340 million tweets sent a day; if you email, the message that you send is one of 144.8 billion email messages sent a day; and if you put videos on *youtube.com*, people are uploading 72 hours of video every *minute*. Obviously this is a data deluge on many fronts. Thus even if our definition is relative, it's fair to talk about Big Data in many situations.

Now how is GI Big Data? Google Earth alone has over 20 petabytes (or about 21 million gigabytes) of data. That may seem small compared to the 2.5 quintillion bytes of data created daily, but processing this amount of data is an amazing achievement. And so, going from the Big Data definition, other data sets are also Big Data because of the current impossibility of processing them in a reasonable amount of time. For instance, 1-meter-resolution aerial photos for the state of Minnesota (86,943 square miles or 225,181 km^2) take 1.05 terabytes of storage.

The main question involved with having so much data is not storing it, which currently depends a lot on cloud computing (networks of computers linked by the Internet), nor is it managing such quantities of data; rather it is the kind of analysis this amount of data makes possible. Most writing about Big Data points to new types of analysis: analyzing the 340 million tweets sent daily to look for instances of the spread of infectious diseases, for example. These are complex and very interesting approaches, generally referred to as data mining. Geovisualization plays a big role in producing representations of this data that are understandable and insightful. Geovisualization approaches can also serve as the basis for predictive modeling. Can established information analysis techniques be grafted on to Big Data? For example, there is the idea of splitting up terabytes of data into smaller megabyte-size tiles to process them on multiple computers using the software employed to analyze small data sets. This is already possible and has been around for a while. For instance, tiles were implemented by ESRI in Arc/Info back in the 1990s. The likeliest scenario is that existing analysis strategies will be supplemented by future innovations.

Issues with Big Data

In a general sense, marshalling Big Data should hold great promise, but even in brief overview we should point to some of the concerns, especially those relevant to GIS.

1. Maintaining data quality is always vexing. Big Data simply involves more complex management of error and accuracy. Adding to this, companies are unwilling to release original data (they mask it using "anonymization" techniques). This means users of the data have to take on faith that the companies have properly checked for and resolved error. Once masking of individual data has been enacted, the impact of this on the quality of location data is very complex.

2. How to link spatio-temporal analysis techniques to Big Data raises a number of complications. Briefly, partitioning Big Data into smaller amounts to be processed using established GIS techniques negates the advantages of Big Data in the first place. Applying spatio-temporal borders during partitioning risks severing important relationships.

3. In typical, non–Big Data GIS, measurement error, subjective bias,

aggregation problems, nomenclature problems, and so on have been corrected for, using techniques that come from classical statistics. How to perform these corrections in a Big Data situation is quite daunting, and some theorists claim that it may be functionally impossible (Weinberger, 2012).

4. Finally, the mainstay of Big Data seems to be the vast amounts of data collected about individuals by companies like Google and Facebook; this involves challenging privacy issues. Conversely, a number of studies (Sweeney, 2002) point to how easy it is to identify individuals if one has access to only a very few markers of their particular behavior. In the United States the status of individual location data seems to be in a state of legal limbo. Data having to do with finances, personal medical data, and data on one's children have some protections, but location data seems unprotected currently. There are a number of moral and ethical concerns involved with the use of geographic data on individuals, and this certainly carries over to the Big Data versions of that information.

In summary, Big Data coming together with GIS has the potential to produce big changes. Big Data processing will benefit from advances in computation, increases in memory, and improvement in connectivity. As the number of sensors deployed in the world mushrooms and the data from those sensors become more accessible to spatial analysis, the potential and pitfalls of Big Data will grow proportionally. With these developments already under way, we would do well to heed the carpenter's credo: "measure twice, cut once."

Summary

The world of online resources has made it possible to *distribute* the maps one makes. Similarly, one can draw from online resources in terms of the data and functionality that comprise that map. Should this be called online GIS or online mapping? That may depend on how the data is displayed and how it is manipulated and controlled. The mushrooming of these technologies and increasing use of mobile devices has led to a vast increase in the amount of location data. Geocoding has advanced to the point where it is possible to speak about a geocoded world, where the data collected by various sources begins to closely reflect things and events in the world. Concerns about the increases in data collection, and its actual and potential uses, have led people to have growing concerns about privacy. Location privacy refers explicitly to the location-related aspects of privacy and their potential to facilitate detailed representations of individuals. Big Data will bring some positive changes to what people do with GIS, but there remain some concerns and big issues to be worked out.

REVIEW QUESTIONS

1. What was a major impetus for developing online GIS?

2. What were the first online GIS able to do?

3. Online GIS enabled innovations. What were they?

4. How is online mapping different from online GIS?

5. Why do online mapping and online GIS have different capabilities?

6. How many times was Google Earth downloaded?

7. How do most people find online GIS data?

8. What organization supports the development of specifications for online mapping?

9. Why is location still so important in the Internet age?

10. What does the term neogeography refer to?

11. What is geocoding?

12. What are some of the issues of data collection, such as from cell phones?

13. What are some of the privacy issues in satellite remote sensing?

14. What are some of the limitations of satellite remote sensing?

15. What kind of data can be collected from Big Data and what are the limitations?

ANSWERS

1. What was a major impetus for developing online GIS?

   People realized that one computer with installed GIS software could be used to produce maps by people at other computers at other offices, in other buildings, other parts of the city or state, even other parts of the world.

2. What were the first online GIS able to do?

   Produce maps on request.

3. Online GIS enabled innovations. What were they?

   Online GIS can integrate data collection, analysis, communication, create new types of operations, and some measure of coordination between offices and remote sites.

4. How is online mapping different from online GIS?

   Online mapping is different from online GIS in two ways. First, online mapping software originates from earth viewer software, which was developed to provide satellite and remote sensing image-based dynamic interactions to users. Second, online mapping enables additional data and choices of symbols to users.

5. Why do online mapping and online GIS have different capabilities?

Online mapping, by design, allows for broader uses and open-ended development. Online GIS applications can have similar capabilities, but that has to be designed and created by individuals or groups with the necessary, and often limited access, to software and servers.

6. How many times was Google Earth downloaded?

Google Earth was downloaded at least one billion times.

7. How do most people find online GIS data?

People who find online GIS data do so using a geoportal.

8. What organization supports the development of specifications for online mapping?

The Open Geospatial Consortium (OGC) supports the development of specifications for online mapping.

9. Why is location still so important in the Internet age?

Location, of things and events, continues to be important because data from multiple sources are aligned or combined according to location.

10. What does the term neogeography refer to?

Neogeography, often shortened to neogeo, reflects the increasing role of geocoded information in the world.

11. What is geocoding?

Geocoding, from a technical point of view, is the process of determining location in a geographic reference system, from other data with a different spatial reference.

12. What are some of the issues of data collection, such as from cell phones?

There are many issues in tracking people. Most have to do with perceived threats to individual privacy.

13. What are some of the privacy issues in satellite remote sensing?

Higher accuracy sensors allow for collection of data that is very detailed.

14. What are some of the limitations of satellite remote sensing?

Most high-accuracy remote sensing technologies will not record details about objects in the shadows of buildings or that are otherwise obscured by natural features.

15. What kind of data can be collected from Big Data and what are the limitations?

All kinds of location data can be collected from Big Data for use in GIS. The limitations are many. Breaking the data into smaller parts to use with a GIS may negate the basic advantages of Big Data. Most common scientific approaches to statistical analysis of data make assumptions that don't usually fit geographic information.

Chapter Readings

Cairncross, F. (2001). *The death of distance: How the communications revolution is changing our lives.* Cambridge, MA: Harvard Business School Press.

Elwood, S., & Leszczynski, A. (2011). Privacy, reconsidered: New representations, data practices, and the geoweb. *GeoJournal, 42*(1), 6–15.

Fischer-Hübner, S., Duquenoy, P., & Hansen, M. (2011). *Privacy and identity management for life.* Berlin: Springer Verlag.

Friedman, T. L. (2006). *The world is flat: A brief history of the twenty-first century* (updated and expanded). New York: Farrar, Straus & Giroux.

Haklay, M., Singleton, A., & Parker, C. (2008). Web mapping 2.0: The neogeography of the Geoweb. *Geography Compass, 2*(6), 2011–2039.

Lanier, J. (2013). *Who owns the future?* New York: Simon & Schuster.

Mayer-Schonberger, V., & Cukier, K. (2013). *Big data: A revolution that will transform how we live, work, and think.* New York: Houghton Mifflin Harcourt.

McLuhan, M., & Powers, B. R. (1989). *The global village: Transformations in world life and media in the 21st century.* Oxford, UK: Oxford University Press.

Pogue, D. (2011). Wrapping up the Apple Location brouhaha. See *http://pogue.blogs.nytimes.com/2011/04/28/wrapping-up-the-apple-location-brouhaha/?pagemode=print.*

Sui, D. Z. (2004). GIS, cartography, and the "third culture": Geographic imaginations in the computer age. *Professional Geographer, 56*(1), 62–72.

Sweeney, L. (2002). Uniqueness of simple demographics in the U.S. population. *International Journal of Uncertainty, Fuzziness and Knowledge-Based Systems, 10*(5), 557–570.

Turner, A. (2006). *Introduction to neogeography.* Sebastopol, CA: O'Reilly Media.

Weingberger, D. (2012). *Too big to know: Rethinking knowledge now that the facts aren't the facts, experts are everywhere, and the smartest person in the room is the room.* New York: Basic Books.

Web Resources

○ A good introduction to Mapserver is available at *www.mapserver.org/introduction.html.*

○ A very insightful overview of Google Earth is at *http://en.wikipedia.org/wiki/Google_Earth.*

○ The Open Geospatial Consortium provides a vast amount of information related to online mapping at *www.opengeospatial.org*.

○ Online mapping has opened countless possibilities for citizen science. A wonderful example of online global land cover validation is at *geo-wiki.org*.

○ James Huggins explains data size measures at *www.jamesshuggins.com/h/tek1/how-big.htm*.

○ The 1-meter resolution aerial photos from Minnesota along with additional information are available at *www.mngeo.state.mn.us/chouse/airphoto/program_details.html*.

○ BroadbandHub provides a short video guide to put Internet data and uses into an analogy with water volumes: *www.youtube.com/watch?v=CsVYID9rMGE*.

○ IBM has a number of interesting resources on Big Data at *www-01.ibm.com/software/data/bigdata*.

○ A well-written discussion of using Big Data for spatial business analysis can be found at *www.forbes.com/sites/tomgroenfeldt/2012/10/15/big-geographic-data-from-tokyo-retailers-to-flood-zones*.

○ An interesting application of Big Data and cartographic visualization to help understand some of the history of globalization is at *http://sappingattention.blogspot.com/2012/11/reading-digital-sources-case-study-in.html*.

○ Professor Shashi Shekhar provides a good overview of spatial Big Data through several presentations at *www.cs.umn.edu/~shekhar/talk/2013/sbd.html*.

EXERCISES

1. Find the Metadata

Find five online data sets for GIS and five online types of GI for web mapping in the same area. The area could be a city, county, state, country, or even a continent. Describe each source of data, how it is made available, and what software is required to make maps with the data.

2. Make an Online Map

Using a local, regional, state, or national online mapping service, create a map of your community. Can you add your own data? Can you modify existing data?

3. **Look into Location Privacy Concerns for Some Web Applications**

Conduct a web search on mobile devices and privacy and identify five types of applications that could raise location privacy concerns. What are the applications? What are the concerns? What can be done to mitigate these concerns?

EXTENDED EXERCISE

4. **GI on the Web**

Overview

Much of the information represented in maps can now be found on the World Wide Web. In addition to presenting reproductions of maps, many websites also offer abilities to produce "customized" maps.

Concepts

Data on the web, some say, is replacing traditional maps. As you work through this exercise think about that statement and about the advantages and disadvantages of maps available on the web.

Part 1: Using Online Data to Make Maps

Step 1

In this part of the exercise you will work with information from several websites. At the U.S. National Atlas site make a map of Minnesota showing either demographic or environmental characteristics.

U.S. National Atlas: *http://nationalatlas.gov/natlas/natlasstart.asp*

Follow the instructions on this webpage and produce a map of Minnesota or the Upper Midwest for an agricultural attribute, such as Soybeans for Beans—1997.

Questions

1. What are the minimum and maximum or first and last legend values?
2. What kind of measurement framework (nominal, ordinal, interval, ratio) is used for the data?

Step 2

Make a map of Minnesota using this site: *http://mapserver.lmic.state.mn.us/landuse*
     Choose a county from this webpage. After choosing a county, generate a table of statistics by choosing the Create Statistics radio button and a county.

3. What is the dominant land use in the county and what is its name?
4. What percentages of the county consist of urban and rural development?
5. Is there any mining in the county?

Scroll to the bottom of the page and use the Create Multicounty Land Use and Cover Statistics link. You will then be at the Minnesota land use and cover statistics page and can prepare statistical summaries for answering the next questions.

6. What Minnesota county has the greatest percentage of forested land?
7. What Minnesota county has the smallest acreage of water?

Step 3

Continue now to look at some demographic data for Minnesota. Go to *www.lmic. state.mn.us/datanetweb/php/census2000/c2000.html.* This brings you to a webpage listing a variety of statistical data available for Minnesota. Choose Demographics from the menu box on the left and then the Census Reports and Mappings link. Under the heading Population Profiles choose Population in 1970, 1980, 1990, and 2000, and then click the Mapping button on the right. Enter a title for the map (e.g., "Minnesota Population in 1970"). Click the Define legend button to make your own legend following this example.

Step 4

Now, let's take a look at an example of a remote sensing application. Go to *http:// earthshots.usgs.gov/Wyperfeld/Wyperfeld* and describe what this imagery shows.

8. What has been happening in Wyperfeld?
9. How many of the 27 seasons were captured by the Landsat satellites? (Hint: read the whole story.)

Part 2: Find Data

In this part of the exercise, you can search for data for any area in the United States you are interested in. Suggested starting points are:

www.mapsonus.com
www.mapquest.com
www.weather.com

Using the site of your choice, print the map you make and answer the following questions on a separate sheet of paper.

10. What is the URL you went to?
11. What kinds of maps are available at this website?
12. Can you interactively create (choose characteristics or attributes, your own legend values, etc.) your own maps at this site?
13. Describe the map you made: What does it show? Does it look like what you would have expected? What is the legend?

Part 3: Take a Bad Map . . . and Make It Better

This is the most creative part of the exercise. You will need to find and print a map that you find "bad" for any number of reasons: the colors, the legend, the theme, the biases (projection, scale, symbols), or whatever. You might want to first go back to Chapter 12 on misuse of maps and map propaganda to see a few examples of "bad" maps. These questions also relate to Campbell's discussion of cartography in Chapter 1.

14. Why is this map horrible?
15. What could be done to improve it?
16. GI on the web, some say, is replacing traditional maps.

After completing this exercise think about whether there are advantages to traditional paper maps that you think web-based maps will not be able to match.

PART 4

Fundamentals of GI Analysis: Understanding the World

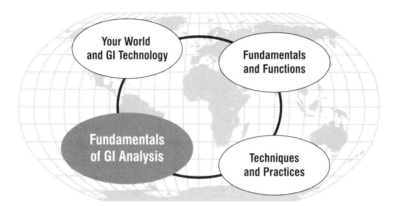

The three chapters in the fourth and final part of *A Primer of GIS* take up using GI to analyze events, ultimately to improve our understanding of the world. The chapters provide examples of the choices one might make and the impact those choices would have on the quality of the analysis.

The central issues in the chapters of this part include operations for extracting information from GIS, data that would then be analyzed (Chapter 15). How to account for error and uncertainty are also covered. Chapter 16, Geostatistics, addresses the statistical and mathematical operations that, when performed on GI, are increasing our ability to get valuable information from the maps and other GI that we now have and will develop.

These chapters are meant to serve as broad introductions to some of what GI analysis is and does. Hopefully, briefly engaging with these topics will help you see the modern potential of GI to enlighten us about so much that would have been harder to grasp in the past.

CHAPTER 15

GI Analysis and GIS

Every day people make spatial choices based on GI: how best to get home from school or work, what is the best spot for fishing, where to go on vacation, where is the most convenient child care, which site is preferred for opening a store. Looking at a map will not necessarily provide you with the best choice. It can help, but reading a map is time-consuming and the map may be inaccurate for your purposes. Sometimes that's obvious, but often it's not. Think of a bicyclist who is looking for the best route from work to home. A street map for cars will only go so far. It doesn't usually show special trails for cyclists and probably doesn't even indicate cycling lanes on the roads. GI analysis can help bring the information together and also leads to the best decision. You often need fuller comparisons and analysis of data to be able to make the best choice. Indeed, GI analysis is becoming more common, due to online and software navigation applications. Analysis is how navigation software determines the possible routes when you just enter your origin and destination. GI analysis also has been and remains a key part of many government and business activities. These navigation applications and professional analyses are similar in concept, but the commercial origin of many online and other software applications means the details of their analyses may be hidden; this restricts consumers. Professional analysis almost always involves considering many details and choosing between possibilities.

In terms of GI, analysis takes many forms. It can involve little more than the comparison of two data sets collected at different times for the same area. Or it can be complex: for example, the analysis of the relationship between existing residences and a proposed highway that will use sound buffers—entailing complex geostatistical analyses that dynamically model the processes of noise being created by different types of traffic through the course of a 24-hour period. A way to think about GI analysis is that it explicitly or implicitly translates or transforms geographic features and events into patterns and processes. A list of intersections (features) is transformed into a route (process) for getting from your origin to your destination. In everyday activities, such analysis is hidden, but without it, finding the way would not happen.

This chapter begins with an overview of the translation or transformation of GI in terms of communication instead of as an operational and functional focus. Considering the communicative role of the analysis helps one get a grasp of the different forms of GI analysis. The chapter next provides an overview of GI analysis types and GIS. This chapter concludes with a discussion of GIS definitions and some examples of how GIS is used for analysis in environmental and urban domains.

Analysis for Communication

Ultimately, all analysis is used for communication. That means that all choices impact what the analysis can be reasonably used for. Given that the historical context, culture, and purpose of any analysis will vary from internal sketches and plans all the way to advertising banners and TV presentations, many choices become implicit or are matters of convention. The choices for analysis involve many of the issues discussed in Chapters 3, 10, and 11. Chapter 12 points to abuses and misuses that can arise from inexpert or unscrupulous decisions, or even choices made on purpose in order to distort. In a nutshell, successful communication relies on appropriate choices connected to geographic and cartographic representation (see Chapter 1 for further information). For example, if an analysis needs to be made of a neighborhood health clinic's accessibility in a large metropolitan area, it will likely be necessary to consider most roads in the area, but not provide a map of the entire region. That's fine for the neighborhood analysis, but using the same map to help patients from other clinics in the region figure out the closest clinic can easily lead to confusion. How choices are made is an important part of every GI analysis.

The choices reflect the conventions of the people preparing the analysis and the cultural values of the people who will be using the results. The issues and choices are numerous and contextual. People familiar with the clinics' locations may leave off information about smaller streets in the cartographic representation that is obvious to them, potentially confusing people who are new to the area or who have never been to the clinic before. But even before the cartographic representation is prepared, the geographic representation will already reflect certain conventions, for example, what counts as a road. The measurements, observations, and relationships of the clinics are strongly influenced by the perceptions, backgrounds, and disciplinary perspectives of the people involved in preparing the analysis.

Issues for GI Analysis

In particular, for GI analysis, three key issues come to mind. These issues often involve trade-offs, not necessarily absolutes or either's and or's, but each issue involves finding different balances between analysis and communication.

Patterns/Processes

One of the most fundamental choices in analysis is deciding how to analyze the relationships involving things and events. Most GIS software only supports the storage of things as patterns. Events can be modeled as processes, with the model aiming to correspond to the dynamics of the events, but with the events being broken down into data captured at particular time points of the event. For example, a traffic jam may be modeled as a process, but the cartographic representation usually relies on a series of "snapshots" created to show the status of the traffic jam at different points in time. The same goes for natural events: the spread of a wildfire modeled as an event may use the same technique to show in an animation how the wild fire spreads.

In other words, while the underlying concepts and geographic representations can take both patterns and processes into account, the cartographic representations usually only show an animation of "snapshots" prepared to show the event's development (see Figure 15.1).

Simple/Complex

GI analysis can range from simple to complex. *Simple analysis* refers to activities involving interpretation or comparisons. *Complex analysis* refers to geostatistical analysis and process modeling. In between the simple and complex types of analysis lies a vast range of analysis types and transformations. For example, a noncomplex GI analysis may consist of simply comparing two data sets of the same area collected at different times, combining a forest-type data set with a soils characteristics data set, or using buffer operations to determine the service area of a proposed bus route change. Some complex analyses may use Monte Carlo simulations to assess which distribution of soil pH values most likely matches the stochastic distribution, use fuzzy-set theories to assess the inaccuracy of boundaries around vegetation types, or rely on variance calculations to help determine the reliability of field data samples.

Many or even most GI analyses will rely on some transformations and an interpretation or comparison. For example, creating a buffer around a factory that is submitting an expansion plan for the facilities is the transformation necessary to combine this buffer with the positions of buildings whose owners need to be informed about the permit application. Many transformations can be linked to a single interpretation or comparison, or each transformation may involve hundreds of data sets that are combined for a single interpretation of how multiple factors influence each other.

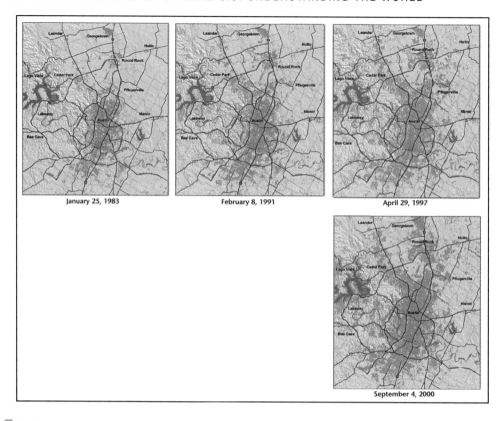

FIGURE 15.1. Growth of Austin, Texas, showing urbanized areas in 1983, 1991, 1997, and 2000. The comparison of snapshots from interpreted remote sensing data can help communicate changes in an area over time.

Accuracy/Reliability

Accuracy refers to the degree of correspondence between data and the actual thing or event. *Reliability* indicates how consistent the data is for certain types of applications. Usually high accuracy means high reliability, but in several circumstances the opposite may be the case.

First, data may be highly accurate, but because of the time between its collection and its analysis, it may no longer be reliable. Aerial photos and satellite imagery can easily become dated and will vary greatly from the actual situation. Second, reliable data may have a low accuracy. For example, data showing the major roads in the United States may be useful for reliably determining how to go from Boston to San Diego, but not accurate enough for determining how far a recycling center in San Diego is located from a highway.

The balance between accuracy and reliability is often a financial issue. Because of the high costs of data collection, often limitations for both accuracy and reliability are acceptable. The key point about this balance is first to clearly describe the data's date of collection along with concerns about possible discrepancies to the actual situation and, second, to take these discrepancies into account when analyzing the data. For example, returning to the clinic example from above, that might mean noting that only roads in a neighborhood were included, thereby making the analysis of little value for the entire region.

Basic GI Analysis Types and Applications

Even just as an overview of the many types of GI analysis and applications, this section gives some insight into pragmatic issues and applications of GI analysis. These GI operations only partially correspond to GIS operations and commands. Depending on the software's analytical capabilities, the analysis in GIS may involve a single command or many commands. Because of the endless permutations of GI analysis operations, a direct match to any single GIS software's set of operations and commands only makes sense for specific applications and domains.

Query

People who mainly have worked with online maps may think of the GI query in terms of interactive maps. In its simplest form, GI querying involves both spatial and attribute aspects that allow a person to select a feature and find out its attributes. This is the contemporary form of spatial analysis first developed by researchers and innovators working on GIS in the 1960s and 1970s. In many applications now, the programming of the query interface can make this a very helpful feature to find out the name of lakes, cities, clinics, restaurants, and so on. For GI analysis using GIS, GI querying is often important in determining the attributes of individual features. Related to the basic "identify" operation are operations for determining the characteristics of features based on their geographical relationship to other features, particular positions, or areas (see Figure 15.2). Another type of query operation makes it possible to select features based on their attribute values. This type of query is a significant analytical tool for choosing features based on combinations of attributes. It is often used after GI is combined to identify particular combinations of attributes.

Combination

One of the most used GI analysis types is widely known by the associated GIS operation: overlay. Overlay in its simplest form involves joining the vector or raster data from two data sets including the attributes. The people who introduced overlay to GIS thought of this type of GI in terms of looking at interactions between two or more transparent maps. The simplicity of this approach was very attractive, as was its chief promoter's (Ian McHarg)

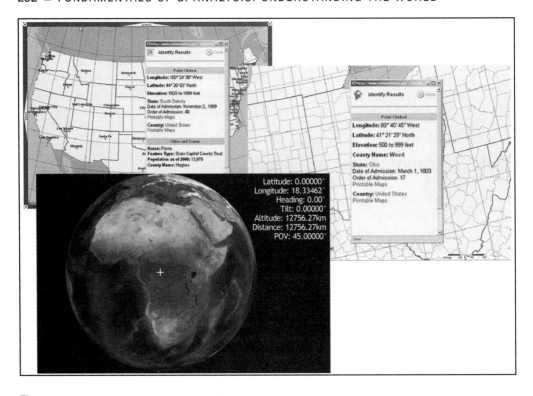

FIGURE 15.2. Different types of identify operations. Identify operations indicate characteristics of a location from the available GI.

emphasis on the overlay operation as a method to incorporate environmental and social concerns along with engineering perspectives in the planning of highways and other large construction projects.

Many people refer to overlay as "integration," but each GIS overlay operation may or may not integrate. As McHarg, Dangermond, and many others from the first generation of GIS developers point out, integrative analysis based on overlay requires the interpretation of the overlay results. Simply combining two (or more) data sets through overlay will only rarely integrate geographically (see Figure 15.3).

The overlay GI analysis operation is prolific due to its ability to take GI from different sources and combine them. That is possible only as long as the GI from the different data sets use the same coordinate system. Additional query GI analysis operations may take place before or after the combination operation (see Table 15.1).

Distance Transformation

The second most used GI type is usually known by its GIS operation equivalent: buffer. The *distance transformation* transforms a feature or area of

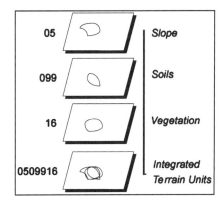

FIGURE 15.3. The combined values of the integrated terrain units must be analyzed to be meaningful. While the methods and details of the operation have changed with the introduction of more capable GIS processing, the basic concepts remain the same.

raster cells into an area based on given distances (see Figure 15.4). One single distance can be used—for example, 100 feet from a well—or multiple distances—for example, 50 m, 150 m, 250 m, and 500 m from a road—for the transformation. The distance transformation GI analysis is often used to show the geographic extent of events (e.g., noise from traffic, leaking of oil tanks into the ground). In these uses the distances correspond to model or assumed values regarding the processes underlying the events. This ability to transform from process to pattern is what makes this GI analysis type most useful.

Neighborhood

Though related to distance transformations, *neighboring* is focused more on establishing what and how features are geographically related. Neighborhood GI analysis usually focuses on using either topology, raster cell neighborhoods, or TIN relationships to analyze geographic relations. For GI analysis of transportation networks, these GI analysis operations are critical to checking and establishing different types of connectivity. In environmental

TABLE 15.1. GI Analysis Types and Related GIS Operations[a]

Type of GI analysis operation	Related GIS operation
Query	Spatial query and attribute query
Combination	Overlay
Distance transformation	Buffer
Neighboring	Connectivity, adjacency, visibility
Rating	Ranking, weighting
Multivariate analysis	Linear factor combinations
Geostatistics	Monte Carlo simulations
	Fuzzy sets
	Variance

[a]Note: GIS operations may not correspond to actual GIS commands.

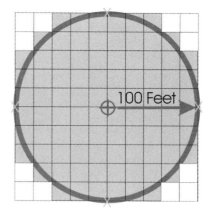

FIGURE 15.4. Distances can be transformed between raster and vector formats with some variations. These transformations can lead to distinct differences in the calculation of the same areas.

applications, neighborhood GI analysis is used for a variety of applications including modeling soil erosion, establishing water runoff patterns, and determining viewsheds (see Figure 15.5).

Rating

Often following or preceding other GI analysis operations, *rating* is used to ordinally rank features based on combinations of attributes or combinations of attributes and locations (see Figure 15.6). It is also often the operation that is a key part of weighting various attributes for decision making. This process is prone to distortions, requiring that great care be taken in determining the weights.

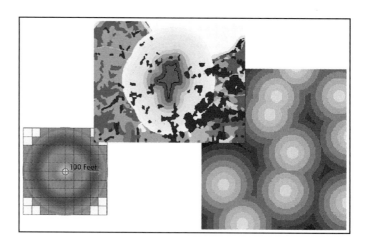

FIGURE 15.5. Examples of neighborhood operations using raster GIS. After overlay, the buffer operation used in these examples is probably the most widely used GIS operation.

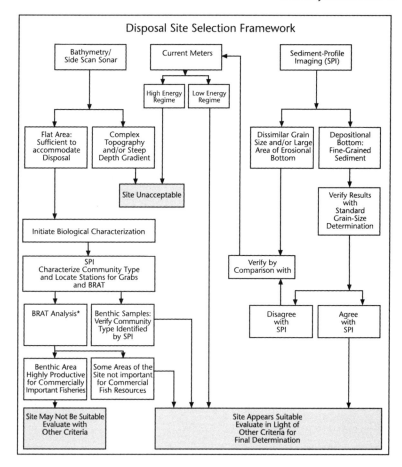

FIGURE 15.6. Criteria and process for selecting a disposal site. These criteria and the order can be implemented in a GIS for very thorough modeling of environmental processes and supporting decision making.

Multivariate Analysis

To avoid some of the problems of weighting, multivariate analysis is a GI analysis operation that facilitates more flexible permutations of attribute combinations and relationships between attributes and geographic positions. The variables are often combined in linear equations that simply provide for the additive combination of attributes and relationships, but multidimensional equations provide ways to consider attributes and relationships in more complex situations.

Geostatistics

The most complex GI analysis operations, and, for many reasons, the most important GI analysis operations, are geostatistical GI analysis operations (covered in Chapter 14). Considering even just three of these rich operations can point out the importance of geostatistics especially for applications that need to go beyond geographic modeling of things as patterns and transforming events into patterns. *Monte Carlo simulations* offer the means to create data that stochastically offers a reliable estimation of the geographic differentiation of continuous characteristics—for example, soil pH, species densities, or transportation costs. *Fuzzy-set theory* is the foundation for a number of GI analysis operations that consider the variability of natural phenomena boundaries and the inaccuracies of data collection. *Variance* measures the difference between repeated measures of the same properties. It is important in assessing the impacts of different geographic aggregation units on the accuracy of data (see Figure 15.7).

GIS in a Nutshell

With many textbooks offering varying introductions to GIS, the purpose of this section is only to show GIS's relationship to GI analysis operations and the role of GIS in representation and communication. Consider first

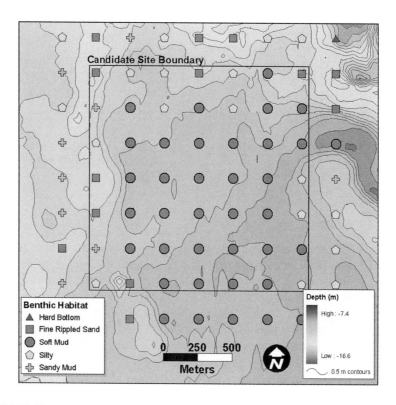

FIGURE 15.7. Site selection for disposal of marine sediments. The characterizations employed in this decision making are used to determine the relative importance of habitats for commercial fishing and suitability for depositing marine sediments.

two common definitions, followed by a third that brings aspects of the two together.

The First Definition

> A system of hardware, software, data, people, organizations, and institutional arrangements for collecting, storing, analyzing, and disseminating information about areas of the earth. (Dueker & Kjerne, 1989, pp. 7–8)

This commonly cited definition focuses on important parts of a system that consists of six components and is used for four generic purposes. It can be used to describe any organized use of GI. GIS definitely involves these components and is used in these four ways. If we consider the terms very broadly—for example, hardware includes notecards and software includes alphabetical filing systems—then even an address list could be considered to be a GIS. In this definition, GIS is used as a single system for all elements of processing GI, beginning with collecting data and ending with making maps and other types of information.

What about the geographic and cartographic representation? Dueker's and Kjerne's definition provides insight into what GIS can be used for. What about how, by whom, and for whom? If you know what the GIS is being used for, then this definition is practical because it lends great flexibility for actually using a GIS in many different ways. It also may be too vague in how it explains the relationships between the components and generic purposes. Are all components equally involved in storage? This seems to be a naïve question, but with this definition standing on its own, as a definition should, you really couldn't tell. This definition is handy, but what most people understand when they rely on this definition is just the surface of GIS; the geographic and cartographic representations are missing.

The Second Definition

> Organized activity by which people measure and represent geographic phenomena then transform these representations into other forms while interacting with social structures. (Chrisman, 1999, p. 13)

Nicholas Chrisman developed this definition as an attempt to address the open questions about Dueker and Kjerne's definition (and many others—see Chapter Readings—that express the same main concepts as their definition). Chrisman's definition focuses on the activities of measuring and representing in the context of social structures. This is the short form of a more involved conceptual model of GIS that consists of a nested set of rings and interactions between the rings.

Chrisman's "shell model" of GIS is broad and inclusive (see Figure 15.8). In its focus on activities, it points to the importance of knowing what any particular GIS is used for. Operations (common GIS processes) and transformations (processes that change the measurement framework) are the emphasis in Chrisman's discussion of how GIS is used. In the redrawn and modified

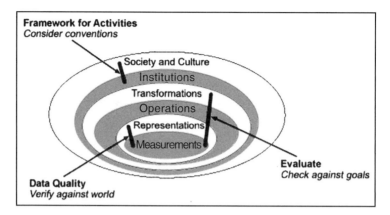

FIGURE 15.8. "Shell model" showing elements of GIS and interactions that occur. The considerations, evaluations, and verifications of creating and using GIS that this figure represents are very important in using GIS.

version of the "shell model" figure, the emphasis is placed on activities that are essential to the successful development and use of any GIS. First, data quality involves verifying measurements and geographic representations in comparison to the corresponding things and events found in the world. Second, the operations and transformations used in any GIS need to undergo an evaluation of each and every use of the GIS. Finally, conventions originating in society, culture, and institutions require consideration to assure that the data accurately correspond to the things and events and the operations and transformations take into account the goals of the GIS.

The Third Definition

A system of computer hardware, software, and procedures designed to support the compiling, storing, retrieving, analyzing, and display of spatially referenced data for addressing planning and management problems. In addition to these technical components, a complete GIS must also include a focus on people, organizations, and standards. *www.extension.umn.edu/distribution/naturalresources/components/DD6097ag.html*

This definition merges parts of the first and second definitions specifying purposes and the importance of "people, organizations, and standards." It starts out with the components of Dueker and Kjerne's definition, adding more specific purposes and rationales and notes the importance of nontechnical elements.

Definitions like these provide more substance for people unfamiliar with GIS, but very interested in using GIS in their organizations. In this definition one of the purposes is "management problems." It also broadens the consideration: this is not just a matter of technology. The people working on the problems, the organizations, and the standards used in developing the GIS must also be considered.

One thing these definitions fail to cover is that "layers" is a common term for describing the organization of data in GIS. Organization in layers is significant because geographic representations of considerably different types are stored in the same GIS. When a GIS uses the same coordinate system, the layers may even be combined. This concept underlies the MPLIS and SDI concepts. Combining two layers can be problematic because of mismatching. Mismatch is revealed as a problem of accuracy. Layers is a visual concept, whereas in a database some layers of data may be stored in a format that is not visual. Nevertheless, the term "layers" helps people understand how different geographic representations are combined.

Examples of GIS Applications

From the plethora of GIS applications, the next sections present a few vignettes that illustrate the roles of geographic and cartographic representation and the significance of accuracy, goals, and conventions in developing GIS applications. Chapter 4 also includes a number of examples.

Environmental and Conservation Applications

Landslide Analysis

In Japan a GIS application was developed to store and manage three-dimensional surface and subsurface data in Akita, Yamagata, and Kanagaw prefectures. The application allows users to construct flexible analysis and examine the effects of preventative structures and plantings, and it can actually simulate landslide movements. Its geographic representation takes various hydrogeological characteristics into account, which can be linked to topographic maps, technical drawings, and orthorectified photos. Organized into layers, the different elements of the geographic representation can be analyzed based on geometric overlaps. Because of the three-dimensional nature of landslides, the cartographic representation supports the use of multiple viewpoints and user-defined visualizations of cut-lines that help users to understand and examine the geological structure. The analyses and visualizations need to be very accurate to meet the goals of users and established conventions of geohazard analysis.

Sea Cliff Erosion

The municipality of Isla Vista, California, near Santa Barbara, has a very high population of students. Apartments overlooking the Pacific Ocean are in high demand, but ongoing erosion has led to several apartments being condemned. Students from the University of California, Santa Barbara, were involved in creating a GIS that mapped details of the cliff edge and the cliff base that could be used in the county GIS. Existing data from the county

GIS was first analyzed and then students went out to collect additional data using GPS and video cameras. To assure that the data was accurate, the cliff data on erosion activity, storm drains, vegetation, and beach access points was collected four times. Using county data, students also analyzed the rate of coastal erosion since 1972. The changes, on average just less than 1 foot of cliff erosion per year, were used to make a prediction of the coastal cliff changes through 2055. The cartographic representations included maps, animations, and interviews. The high accuracy of the data points to the importance of having robust scientific data to fulfill conventions for data that will be used by the municipality and county in making decisions (see Figure 15.9).

Urban Applications

Improving Land Administration

In many countries around the world, cadastral records, planning documents, and planning maps needed for land administration are spread among various agencies. Collecting and comparing documents can require a great deal of time, and problems and conflicts often arise. Many GIS applications have been developed to assist governments with land administration. These applications must take the existing geographic representations into account, but also determine what are common agreements and disagreements between the government agencies. By developing a geographic representation for the GIS that helps facilitate interactions and flags possible problems for resolution, the GIS can greatly improve government land administration. Developing cartographic representations that accurately indicate different planning

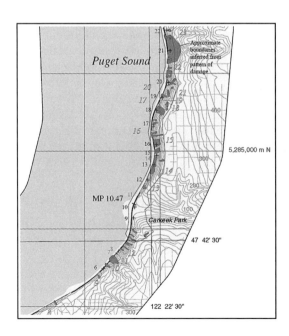

FIGURE 15.9. Section of a map produced with a GIS-based analysis of landslide hazards on Puget Sound. Maps are often used as the base graphic for presenting results of GIS analysis to help communication.

zones and ownerships, and that help administrators understand the problems, are critical to the success of these applications. Through this support and using information technology to improve the accuracy and speed of processing, land administration can make important decisions much quicker.

Modeling Urban Growth

The rapid increase in urban populations around the world (more than 50% of the world's population now live in cities) leads to a number of health, social, and environmental problems. To help administrators and politicians develop a better understanding of this growth, research projects modeling the process of urban growth have helped predict the future growth of areas. These studies require very careful consideration of geographic representation issues. For instance, people will tend to migrate to areas with good transportation to employment possibilities, but these areas often have the highest rents. Many people will look for nearby alternatives that help them save on housing costs, but are still close to transportation. The development of an area can accumulate and lead to very fast growth as more people choose a place to live, more transportation is provided, more people come, and housing costs increase. After a certain point people will move to new nearby locations, starting the cycle over again in a new place. The geographic representation of these interactions requires detailed modeling of the numerous political, social, and economic factors that influence the process of urban growth. The detail of the geographic representation is also important for assessing the accuracy of the model and assuring that the goals of the modeling are met. The cartographic representation of these models needs to take conventions for mapping an area into account to assure that people can understand the results of a model. See Plate 12 for an example of a field model.

Summary

GI analysis improves spatial decision making and communication. It takes many forms, ranging from the comparison of data sets to dynamic models of interactions. The choices made in analysis reflect the same conventions involved in any type of geographic representation or cartographic representation. Three issues are especially significant for GI analysis: (1) choosing patterns or processes, (2) applying simple or complex analysis techniques, and (3) determining the appropriate balance between accuracy and reliability. Basic GI analysis techniques include analysis, combination (overlays), and distance transformation (buffers). More complicated techniques include neighborhood analysis and rating. Multivariate analysis is a more complex form and geostatistics is the most complex form of analysis.

GI analysis relies on GIS, which is also used for cartographic presentation. GIS is used for every kind of GI analysis and has become as significant as the microscope in changing how people analyze the world.

REVIEW QUESTIONS

1. What should GI analysis consider so as to communicate the desired intent?

2. What is the difference between patterns and processes?

3. Why distinguish between simple and complex types of analysis?

4. Which GI analysis type is more common, buffers or overlays?

5. Is there a trade-off between accuracy and reliability?

6. How is Chrisman's definition of GI different from the "input–process–output" definition?

7. What are the elements of Chrisman's GIS definition?

8. How does the limited ability of GIS to consider process constrain considerations of events?

9. What is the difference between a translation and a transformation?

10. What is a common application of buffers?

ANSWERS

1. What should GI analysis consider so as to communicate the desired intent?

   GI analyses should consider the history, culture, and purposes along with issues of geographic and cartographic representation.

2. What is the difference between patterns and processes?

   Patterns are geographic representations that portray a static geographic situation or a snapshot of an event. *Processes* represent the geographic interactions dynamically.

3. Why distinguish between simple and complex types of analysis?

   This distinction helps one to grapple with analysis applications that make more of the choices underlying the analysis visible.

4. Which GI analysis type is more common, buffers or overlays?

   Neither. Both are of great importance generally. Specific disciplines or applications may use one or the other more often, but across the board both are very important.

5. Is there a trade-off between accuracy and reliability?

   Generally not, but in cases where positional and temporal accuracy are both involved, there may be a trade-off between lower positional accuracy and greater temporal accuracy or vice-versa.

6. **How is Chrisman's definition of GI different from the "input–process–output" definition?**

   Chrisman's definition accounts for different interactions and context issues.

7. **What are the elements of Chrisman's GIS definition?**

   The four "activities" in Chrisman's GIS definition are making measurements of geographic phenomena and processes, making representations of what was measured, performing further operations, and making transformations to other representational systems.

8. **How does the limited ability of GIS to consider process constrain considerations of events?**

   Events in most GIS need to be geographically represented as things. A set of things ordered by the time of their observation can be used to create an animation, but this dynamic visualization does not necessarily correspond to the actual process.

9. **What is the difference between a translation and a transformation?**

   Translations are done by humans, usually working with GI on computers, but sometimes working only with maps and other printed material. Transformations are done by GIS software to produce new GI from existing GI, for example, a buffer to generate the extent of an animal's biotope from the site of its nesting.

10. **What is a common application of buffers?**

    Buffers are commonly used to create a zone that corresponds to the effects of a process. For example, buffers offer a crude way to represent the spread of noise from vehicles, airplanes taking off and landing, or pollution emissions from a smokestack or outlet pipe.

Chapter Readings

Chrisman, N. R. (1999). What does "GIS" mean? *Transactions in GIS, 3*(2), 175–186.

Dangermond, J. (1979). A case study of the Zulia Regional Planning Study, describing work completed. In G. Dutton (Ed.), *Urban, regional and state applications* (Vol. 3, pp. 35–62). Cambridge, MA: Harvard University Press.

Dueker, K. J., & Kjerne, D. (1989). *Multipurpose cadastre: Terms and definitions.* Falls Church, VA: American Society for Photogrammetry and Remote Sensing.

Goodchild, M. F. (1978). Statistical aspects of the polygon overlay problem. In *Harvard papers on GIS: First International Advanced Study Symposium on Topological Data Structures for Geographical Information Systems.* Cambridge, MA: Harvard University Press.

Haklay, M. (2004). Map calculus in GIS: A proposal and demonstration. *International Journal of Geographical Information Science, 18*(1), 107–125.

MacDougall, E. B. (1975). The accuracy of map overlays. *Landscape Planning, 2,* 25–30.

McHarg, I. (1969). *Design with nature.* New York: Natural History Press.

O'Sullivan, D., & Unwin, D. J. (2003). *Geographic information analysis*. New York: Wiley.

Tomlin, C. D. (1990). *Geographic information systems and cartographic modeling*. Englewood Cliffs, NJ: Prentice Hall.

Veregin, H. (1995). Developing and testing of an error propagation model for GIS overlay operations. *International Journal of Geographical Information Systems, 9*(6), 595–619.

Web Resources

↻ For a broad overview of GIS analytical use (with an emphasis on caving), see *http://web.archive.org/web/20060924125931/http://rockyweb.cr.usgs.gov/outreach/articles/nss_gis_article.pdf*.

↻ Examples of the application of GIS-based GAP analysis are available online at *http://gap.uidaho.edu/index.php/presentations*.

↻ Applications from around the world showing uses of GIS and remote sensing in urban analysis are available online at *http://web.mit.edu/urbanupgrading/upgrading/case-examples/index.html*.

↻ Archaeological applications involving GIS-based analysis are available online at *www.informatics.org/france/france.html*.

EXERCISE

Describing and Evaluating a GIS Application

Concepts

GIS involves the organization of many different aspects into a system. These aspects are interdependent. In this exercise, you will discuss the aspects of a GIS in terms of the definitions from Chapter 13.

Objectives

Write an essay that systematically describes the components, issues, and activities of a GIS application.

Activities

Using Chrisman's definition, organize the outline for an essay that includes the components, issues, and activities described in the text. Make sure to address the accuracy, conventions, and goals of the application you choose.

CHAPTER 16

Geostatistics

Geostatistics involves both the most complex and the most important GI analysis operations in part because of the broad uses of geostatistical analyses operations, but perhaps even more due to the underlying power of the mathematical analysis of GI. For many people geostatistics is far easier to grasp than abstract mathematics because its mathematics are tied to actual things and events.

This chapter provides an introduction to the concepts of geostatistics. First, we examine the concepts by themselves, followed by discussion of some applications. Geostatistical applications run the gamut of statistics, applying techniques and concepts from classical probabilistic statistics to Bayesian-based statistical analysis. In many environments these applications are only the first cut of more detailed analysis required for assessing geographic patterns, processes, and relationships.

Patterns Indicate Processes

In geostatistics, patterns express evidence of spatial processes (see Figure 16.1). At first you may think of pattern in a visual sense—for example, a map showing temperatures across the United States. But, in geostatistics, the visible pattern points to the possibility of underlying relationships that can be expressed mathematically. Geostatistics relies on measurements instead of cartographic representations for analysis, although it later uses cartographic visualization extensively. An important distinction from "classical" statistics lies in how spatial correlations are considered in geostatistical analysis. The geographic distribution of temperatures helps illustrate the important

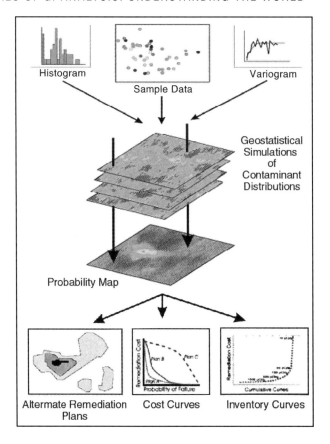

FIGURE 16.1. Example of environmental data sampling showing how data is combined in geostatistical analysis.

difference. Following fundamental physical laws, the temperatures of any body will tend toward equilibrium, that is, temperatures that start out very different will rise or fall toward an equilibrium point, somewhere around the average. But clearly weather, with cold fronts, warm fronts, winds, human influences, jet streams, clouds, and many other characteristics, is dynamic. Temperature is just one indicator at a particular place of the weather. And the temperatures measured at various places in an area at a particular point in time form a pattern that indicates relationships between fronts, jet stream, and so on. By considering temperature measures made at other points in time in the area, the comparison of different temperature measures leads to some understanding of the underlying processes.

These comparisons can be made visually, but are then plagued by uncertainty arising in interpretation. Scale is an important factor: if we consider temperature differences in the entire United States, then it would be difficult to tell clearly what the differences in temperature are between Philadelphia and Baltimore. Visual symbols, especially color, may be easily misinterpreted. If one is using a large number of temperature maps and if the level of detail is significant, the comparisons will also run into difficulties when one tries to compare temperature changes over a year, or even over a season. A person might be able to interpret differences between five maps of the United States, but comparing 500 temperature maps would be impossible for almost anybody.

Geostatistics usually works directly with underlying measurements (see Figure 16.2), rather than the higher-level visual representations. By working directly with the measurements, one can examine the relationships between places and the processes under study in more detail regardless of scale, regardless of different visual interpretations of color, and regardless of the number of maps. In other words, even if statistical operations involve only simple comparisons, the mathematical operations help one grasp the complexity of mapping more than visual techniques would. Additionally, because measurements are being used directly, this elucidates issues of accuracy and the validity of measurements at the points where measurements were made and, by extension, the places in between.

Patterns indicate characteristics or phases of processes, but relationships between patterns and processes are complex. Two similar terms that should not be confused are correspondence and correlation. *Correspondence* means that you are likely to find A with B. That might be what you intuit, for example, when you try to guess how many people would need to be in a room to have a 50-50 chance of having somebody in that group share your birthday. If you think 100, that's too many; 50 is also too many. Statisticians have shown that over and over again you only need 23 people to have that

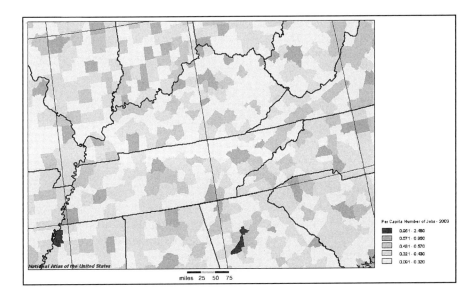

FIGURE 16.2. *nationalmap.gov* offers an interactive mapping application—here, the per capita number of jobs for a portion of the country is shown.

50-50 chance. *Correlation* is the term in statistics that refers to the strength of the relationship between two data sets, for example, between A and B. As an example, does warm summer weather mean larger monthly rainfall? Other factors may need to be considered for more accurate insights into weather processes. While temperature is important, it is only one indicator of weather. It also varies considerably during the day according to the location's proximity to large water bodies, forests, and so on. Geostatistics helps us to understand these relationships, although often the measurements of the statistics should be considered carefully to gain a better understanding of the processes.

You also need to be aware of the complexity of establishing, creating, and validating measurements. When measurements are collected on the environmental characteristics of things and events, they are usually collected at distinct places. How these places are determined is a crucial question for assessing the reliability and validity of the measures. The sampling distribution must be carefully decided on based on an analysis of known properties and processes, or through a simulation of the patterns and processes to be studied. The types of instruments and the types of measurements are another important point to consider. If different instruments are used in different places, the measurements may be significantly different. If the exact differences can be ascertained, it may be possible to transform the measurements to a common reference system. For example, measurements of temperature in Fahrenheit and in Celsius need to be converted to one system before they can be compared. Both the establishment and the creation of measurements have significant impacts on the validation of measurements. Validation is an important separate part of data collection, necessary to assure the internal validity of measurements made at the same place over time and the external validity of the measurement to other places, other instruments, and other types of observing and recording the same types of measurements.

Spatial Autocorrelation

Geostatistics is powerful, but easily prone to great errors through simple misrepresentations or because of great differences between most types of statistical data. The most significant problems are that geographic data sets are not random, nor samples. They are spatially dependent, or spatially auto-correlated. Spatial autocorrelation indicates how spatial dependency varies between a sample and its neighbors. Its most famous expression is found in Waldo Tobler's First Law of Geography, "I invoke the first law of geography: everything is related to everything else, but near things are more related than distant things" (Tobler, 1968). Remember, a basic principle of "classical" statistics is that data is randomly distributed—for example, the ages of people in a city have nothing to do with the city itself. In other words, "classical" statistics doesn't consider the impact of spatial autocorrelation. Obviously, as geographers can explain, all geographical data is affected, sometimes even determined, by its location. Amphibians flourish near water bodies, schools are located near where children live, stores are accessible to shoppers, and traffic jams occur on busy roads. The concept of Tobler's first law may seem blatantly obvious to geographically minded people, yet it is completely at odds with the principle of "classical" statistics. ("Classical," however, often distinguishes much of statistics from Bayesian statistics.) Geographical things and events can sometimes be more complicated. For example, infectious diseases can spread by plane travel across oceans. Geographers call this "jumping scale" because the infection, which normally spreads at a local scale, "jumps" to a different position on the globe and then becomes active at the local scale again in the area where the plane lands.

Following the concept of spatial autocorrelation, it is easy to grasp the idea that samples—for example, measurements of temperature or anything else—cannot reflect a completely uninfluenced area, where the measured characteristics follow a random distribution. For example, let's assume that most frogs prefer a partly wet habitat. If a researcher is determining the impact of household pesticide use on the local frog population, it makes the most sense to collect frogs near water bodies in low-lying wet areas. However, that may or may not be the places where people live and apply pesticides. The relationship of residential location and wetlands needs to be considered in collecting the frogs. The basic principle applies to any type of geostatistical data collection: all sampling and collection of data needs to take account of the underlying processes and factors that influence the processes. The complexity of relating sampling and data collection to things and events makes geostatistics very complex, yet, when properly and reliably done, very reliable. Understanding spatial autocorrelation as it relates to data is important to understand the significance of the spatial statistics and avoid erroneous conclusions from the analysis. A semivariogram that shows x as the distance between locations and y as the strength of the relationship demonstrates the strength of the relationship at different distances.

The Ecological Fallacy

A significant mistake easily made when working with geostatistics is based on what is known as the "ecological fallacy." The *ecological fallacy* is the assumption that the statistical relationship observed at one level of aggregation holds at a more detailed level. A well-known example of this occurs when people look at statistics of election results. In most U.S. states in the 2004 election, the majority of voters voted for George W. Bush, but most people in the cities voted for John Kerry—if you only counted the votes from a city and assumed it applies for the whole state, you would be in error. Take another example: there may be a strong relationship between the number of zebra mussels in lakes and the number of recreational boaters in a state, but other local and regional factors in a state may be more significant in explaining the relationship. The strong statistical significance at the state level may be hiding the factors that actually lead to large numbers of zebra mussels. (Figure 16.3 shows some possible interactions at a global scale involving population and health care.) Something completely different may be causing the high number of zebra mussels. Perhaps it is the size of lake, the number of boats using multiple lakes, or the number of below-freezing days in the area. In every case of geostatistics, the relationship between the aggregation units and the things and events being studied must be carefully examined.

Modifiable Areal Unit Problem

The modifiable areal unit problem (MAUP) is a special instance of the ecological fallacy that results when data collected at a more detailed level of aggregation—for example, census blocks, counties, or biotopes—are aggregated to less-detailed levels of aggregation—for example, census tracts, states, or watersheds. The aggregation units may be arbitrary in terms of the things and events being studied, but the aggregation units used in collecting or collating the data will affect the statistics based on this data. The consequences of the MAUP can be significant. Although the assignment of counties to states is political, statistics show that a switch of one northern Florida county to Georgia or Alabama would have produced a different outcome in the 2000 U.S. presidential election.

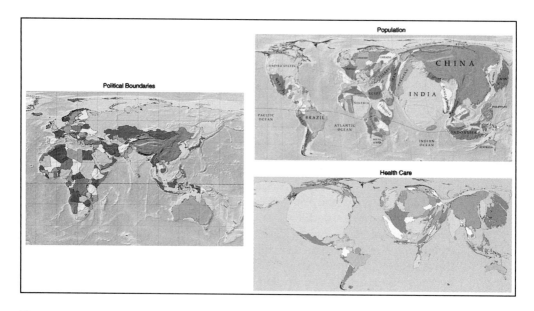

FIGURE 16.3. Cartograms show how other geographical characteristics influence mapped attributes, issues that geostatistics can take into account.

Terrain Analysis

Terrain analysis is an important application domain of geostatistics for a number of disciplines and professions. Civil engineers rely on terrain analysis when planning the construction of large structures, cell phone companies use terrain analysis to plan the siting of antennas, city planners rely on terrain analysis to assess the impacts of new buildings on the landscape, and the military uses terrain analysis for planning and preparing missions. The list could go on.

Because of the breadth of applications and the number of operations and variables, the role of geostatistics in terrain analysis is hard to define. To begin, you can distinguish types of terrain analysis by the role of visual interpretation in the application. Some applications begin with a visual analysis of field data, maps, aerial photographs, remote sensing images, or various combinations of these materials. Other applications may start out with geostatistical analysis or rely on geostatistics to analyze the materials discovered during a project.

For example, many archaeologists use terrain analysis by beginning with a visual inspection of aerial photographs or remote sensing data to see if traces of previous habitation or structures are visible. After comparing these materials with previous archaeological projects and documents, the researchers often look to geostatistical techniques as a means of examining multiple sources of data and evaluating characteristics in the data sources for relationships that can help with understanding the previous cultures in an area. A common application of geostatistics begins with analyzing previous habitation patterns to establish geographical relationships. Among these relationships are distances of settlements to water bodies, trails and roads, and areas with different types of agriculture. These relationships can be quantified as measured distances and used to establish areas of potential settlements, agriculture, and habitation. These areas can be combined by using an overlay operation or geostatistically analyzed to assess the combined potential that any particular area was used by the past culture. This type of application is helpful in planning potential archaeological digs. It can be extended and detailed in a variety of ways.

The use of geostatistics in archeological analysis such as this example may involve a simple factor analysis of the geographical relationships. A *factor analysis* is an explanatory analysis technique that lacks an assumption that the factors are independent. The factor analysis is therefore prone to substantial variability arising in the determination of factors and the assignment of weights (see Figure 16.4). Why, for example, does the distance to water bodies have a weight of 0.2 while the proximity to trails and roads only has a weight of 0.1? The sense of these weights is not transparent, but takes into account the specialized knowledge of the archaeologist. The explanation may help, but if the factors and weights are badly chosen, they may lead to misleading results (see Figure 16.5).

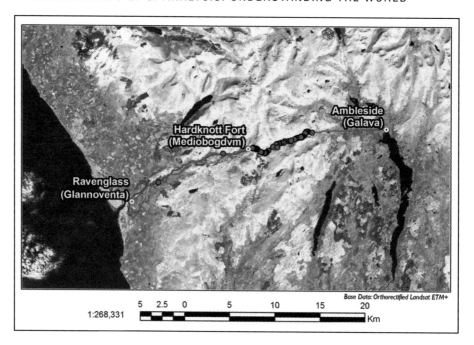

FIGURE 16.4. Archaeological least-cost model of the possible path of a Roman road in England. Slope is considered to be the primary determinant of the possible path of the road.

Types of Terrain Analysis

Another way to distinguish different types of terrain analysis is to distinguish between two-dimensional and three-dimensional terrain analysis. While there is a fuzzy boundary between the two and this division means separating the visual presentation of results from the analysis, the distinction helps one to get a basic grasp on fundamental types of terrain analysis.

Two-dimensional terrain analysis, or 2-D, usually relies on the use of

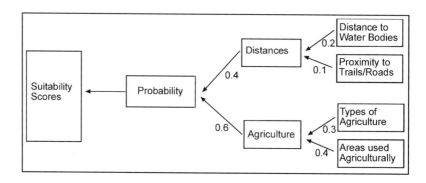

FIGURE 16.5. Weighted factors are often used to determine suitability. The weights adjust the factors for subjective values.

raster data to analyze relationships based on location. A common 2-D raster terrain analysis is visibility analysis, called "viewshed analysis." This operation assesses the area of cells that can be seen from an origin, a raster cell, by comparing the elevation of the neighboring cell with the origin and going outward to other cells if the cell in question is at an elevation equal to or below the elevation of the origin cell. This type of terrain analysis is important in environmental mediation, planning, and the location of transmission towers. It also can be used as part of more complex terrain analysis—for example, considering the impact of large-scale landscape changes on snowmelt processes.

More complex raster-based terrain analysis is used for countless environmental applications. Watershed processes, erosion, and sediment-yield analyses often use terrain analysis as part of more comprehensive analysis that takes account of dynamic processes that are ill-suited for factor analysis. A sediment-yield analysis, for example, can consider soil types, vegetation, geology, maintenance practices, water absorption capacity, and numerous other factors as part of a dynamic principal component analysis that accounts for the amount of precipitation and duration as key variables. Principal component analysis aids the simplification of complex statistical relationships through the identification of independent and uncorrelated variables.

Vector-based terrain analysis is used less frequently because the GI representation of vectors in commercial GIS produces sharp boundaries that generally do not reflect the field nature of environmental things and events. It does find many applications in detailed engineering work—for example, in the calculation of cuts and fills in construction projects, or the determination of elevations.

Three-dimensional terrain analysis (3-D) is most commonly performed with TIN-based terrain analysis or, less frequently, with 3-D rasters, called "voxels." TIN is used in analysis of surficial processes—for example, determining erosion or assessing watershed runoff. TIN-based terrain analysis often uses a digital elevation model (DEM) and water body or road network data. The geostatistics involved in DEMs are commonly linear equations, but they can involve principal component analysis, fractals, and shortest path analysis. Voxels are used in specialized applications to analyze relationships in the atmosphere or in the ground, including the diffusion of aerosols from factories and the pollution of groundwater.

Chi-Square Analysis

Chi-square analysis is a straightforward statistical technique that can be used to evaluate the validity of a hypothesis. Although comparatively simple, it is widely used to examine the relationship between two variables in a cause-and-effect relationship. Because it is reasonably straightforward, it is widely used for exploring a number of questions. For example, does higher soil pH lead to less healthy plants? Does proximity to bus stops lead to an increase of people using public transit? How strong is the relationship between crop types and water resources? It may need to be followed up by more exacting geostatistical study, but chi-square analysis often serves as a key starting point for testing and validating questions. Chi-square analysis compares an idealized random distribution with existing or projected distributions. The random distribution of variables for chi-square analysis that measure characteristics of things or events is called the "normal distribution," or what one would expect if there is no relationship between the variables. The existing distribution is called the "expected distribution." We can go through an example, step by step, to get a better idea of how chi-square analysis works.

1. Create an Observed Frequency Table and Examine Relationships

The first step is to organize the data into a contingency table where the rows depict one aspect of the independent variable (considered the "causal" factor of the relationship) and the columns depict the dependent variable (the "effect" from the independent variable). For this example, we will consider how elevation affects snowfall. Our hypothesis is that more snow falls at higher elevations. We have the data from 133 observations at elevations between 500 and 4,500 m and the yearly average amount of snow, which ranges from 0 to 534 cm. In Table 16.1, the observed frequencies table, the rows indicate elevations, classified into three groups, and the columns indicate the snowfall averages, classified into three ordinal categories. The individual cells give the number of cases that fit into the combination of elevation and snow—for example, 10 observations at elevations over 2,000 m have, on average, more than 60 cm of snow yearly. Looking over all the cells, one sees a marked tendency in the observations that suggests higher elevations receive more snow than lower elevations do. The chi-square statistic is used to test the relationship between the two qualitative variables.

2. Formulate a Test Statement

Each distribution accounts for both independent and dependent variables. The test statement expresses the relationship we think we see in the data in a

TABLE 16.1. Observed Frequencies

| Elevation | Yearly snowfall | | | |
	Low (0–20 cm)	Medium (20–60 cm)	High (60 + cm)	Totals
500–1000 m	41	19	10	70
1000–2000 m	22	15	6	43
2000 + m	2	8	10	20
	65	42	26	133

more specific manner. The established approach for creating this *null hypothesis* is that it states that the variables are not associated. If the chi-square statistic disproves the null hypothesis, then the opposite is proven, namely, that more snow falls at higher elevations. In statistical terminology, H0 refers to the null hypothesis; H1 refers to the alternative hypothesis.

We use the chi-square statistic to determine the difference between the actual observations and what we would expect if the observations followed our null hypothesis. We need to create a second table based on the assumption that the null hypothesis is correct and then compare the two tables. The values for the second table, called the "expected frequencies," are calculated by using the row and column totals (Table 16.2). First, calculate the expected probability that the snowfall is low by dividing the total number of observations of low snowfall by the total number of observations (round the results to three significant digits):

$$\text{Expected probability (low snowfall)} = 65/133 = 0.489$$

Second, calculate the expected probability that the observations are between 500 and 1,000 m:

$$\text{Expected probability (500–1,000 m elevation)} = 70/133 = 0.526$$

Since the null hypothesis assumes they are independent variables, calculate the combined expected probability by multiplying the two expected probabilities together:

$$\text{Combined expected probability} = 0.489 \times 0.526 = 0.162$$

Then multiply the total number of observations (133) by the combined expected probability to determine the expected frequency of low snowfall at 500–1,000 m elevation:

$$\text{Expected frequency} = 133 \times 0.257 = 34.209$$

TABLE 16.2. Expected Frequencies

Elevation	Yearly snowfall			
	Low (0–20 cm)	Medium (20–60 cm)	High (60+ cm)	Totals
500–1000 m	34.209	22.107	13.648	70
1000–2000 m	21.006	13.575	8.379	43
2000+ m	9.709	6.251	3.79	20
	65	42	26	133

TABLE 16.3. (Observed – Expected)²

	Yearly snowfall		
Elevation	Low (0–20 cm)	Medium (20–60 cm)	High (60+ cm)
500–1000 m	46.118	9.653	13.308
1000–2000 m	0.988	2.031	5.660
2000+ m	59.429	3.059	38.564

Repeat these four steps for each relationship (the other eight cells) between snowfall and elevation to complete the expected frequencies table.

Remember that the expected counts in each table square are not the actual observations, but are based on the assumption that there is no relationship between the two variables. If we even visually compare the two tables, we can see differences suggesting that the idea that more snow falls at higher elevations is probably right and that the null hypothesis will be disproven by the chi-square statistic. To calculate the chi-square statistic and have more certainty than our visual inspection, we need to first square the differences between the observed and the expected counts (Table 16.3):

$$\text{Squared difference (low snowfall}/500–1,000 \text{ m elevation)}$$
$$= 41 – 34.209 = 6.791^2 = 46.118$$

This result, however, is dependent on the number of observations. The final calculation of the chi-square test standardizes the calculation regardless of the number of observations (Table 16.4).

The sum of the cell values, 42.206, is the final chi-square statistic. We now need to consider how the number of variables influences the null hypothesis acceptance or rejection, something called "degrees of freedom" in statistics. You have already seen how the number of observations could affect the statistic and was taken account of. The number of variables can have an effect as well. The degrees of freedom is calculated by multiplying the number of rows—1—by the number of columns—1—in the observed or expected tables.

TABLE 16.4. (Observed – Expected)²/Expected

	Yearly snowfall			
Elevation	Low (0–20 cm)	Medium (20–60 cm)	High (60+ cm)	Totals
500–1000 m	1.348	2.086	3.379	6.813
1000–2000 m	2.195	3.397	5.504	11.097
2000+ m	4.750	7.378	12.168	24.296
Total	8.294	12.861	21.051	42.206

$$\text{Degrees of freedom (df)} = (3 - 1) \times (3 - 1) = 4$$

Using a probability table showing degrees of confidence and confidence level (alpha), we can establish the number of times out of 100 that would be exceeded if the null hypothesis were true. In this case, with four degrees of freedom and at a confidence level of 0.05, the probability is 9.49. The value from the chi-square (42.206) is significantly higher. This means that the probability of getting a chi-square value this high is much lower. In other words, we can now pretty certainly say that we should reject the null hypothesis and note that the relationship between the amount of snowfall and elevation has been statistically supported.

From this point, we could move on to other statistical tests to look at specific relationships between snowfall and elevation. We might also want to assess our chi-square statistic, comparing it with other data and trying to make our statistic more robust. For example, we only used 20 observations in high-elevation areas. Could we get more observations? Should we find out where the observations are located to determine if there are any biases? These questions reflect concerns that need to be expressed when using this data, as the chi-square statistic itself fails to account for general or specific geographical relationships.

Spatial Interpolation

A problem for many applications of geostatistics is that data is available for selected points in an area, but not for the entire area. A common example is soil pH, which can be collected only at distinct points using testing equipment. Basically, any ground, water, or air property is based on measurements recorded at individual points, making spatial interpolation a very useful transformation.

A number of factors influence spatial interpolation. Most significantly, the choice of technique will have great impacts. A number of statistical and deterministic techniques, from simple to complex, are available. Choosing one depends on a number of factors and experiences. The two techniques that are often used are the local spatial average and the inverse-distance-weighted spatial average. Splines and multiquadric analysis and TINs are also widely used, but this section focuses solely on spatial averaging.

Local spatial average interpolation considers multiple sample point values when determining the interpolation. The interpolation can specify the maximum area for calculating the average, the number of points, and the maximum distance a point can be considered in the average. Closely and regularly spaced sample points can be interpolated with little overlap or blank spaces. If the sample points are irregularly spaced or far apart, the interpolation may end up with gaps and deviate from the actual values considerably. Increasing the area for averaging may help, but can also turn the interpolation into averages that bear little resemblance to the actual area and the sample points. Just considering the nearest-neighbor sample points is more reliable, but can produce inaccurate interpolations if the points are far apart.

In contrast, the inverse-distance-weighted spatial average technique offers some possibilities to address the weaknesses of local spatial average interpolation. This technique gives nearby points more significance in calculating the interpolation than more distant points. The weight is proportional to the inverse distance between the origin point and the sample point to be interpolated. In other words, as distance decreases, the significance of the sample point increases. This weighting takes account of autocorrelation.

The weights (see Figure 16.6) vary by distance. The distance weighting can be altered. In the figure, it is close to a linear relationship between the distance and the weight. Other than the weighting itself, the resolution of the grid being produced by the interpolation and the order of choosing points for evaluation can have subtle effects on the results. More specific control is offered through kriging.

Summary

Geostatistics offers the most complex and important GI analysis operations because of the underlying power of its mathematically oriented geographic representations. In geostatics the patterns recorded as GI represent spatial processes. The mathematical geographic representation is well suited for analysis of the spatial relationships among complex and diverse indicators that are beyond direct human interpretation—for example, the analysis of election results from the over 3,200 U.S. counties. Geostatistics works directly with measurements instead of with cartographic representations. Work with geostatics therefore requires paying attention to the complexity of establishing, creating, and validating measurements.

Because of the effects that aggregation to different units—for example, counties to states—has on measurements, spatial autocorrelation is an

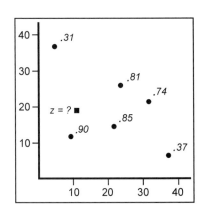

FIGURE 16.6. Example of weights assigned to elevation (z) based on distance points being evaluated.

| IN DEPTH | **Kriging** |

The arbitrariness of distance-weighting functions used in spatial interpolation can be addressed by specifying the general form of the function and using point sample data to determine the exact form. Polynomial equations make it possible to see clear trends in the data, rather than to fit a rigid structure to the data with inverse-distance-weighting interpolation.

Kriging is based on a theory of regionalized variables, which means that distinct neighborhoods have their own variables. This leads to the calculations being optimized for neighborhoods rather than for the entire area. Kriging uses inverse-distance-weighting interpolation for each neighborhood, which can also be varied in a neighborhood. This involves three steps: (1) assessing the spatial variation of the sample points, (2) summarizing the spatial variation, and (3) using the spatial variation model to determine interpolation weights. The mathematics are summarized by a number of authors. Plate 13 shows an example of kriging.

---

important issue to keep in mind. Basically, the principle of spatial correlation is that things are more similar to near things than to faraway things. Classical statistics assumes, however, that things, no matter where they are, are randomly related to other things.

The modifiable areal unit problem (MAUP) is the assumption that a relationship observed at one level of aggregation holds at another, more detailed, level—for example, that the majority of voters in a city voted for the candidate who won the state election simply because the candidate won the state election.

Geostatistics are applied in many areas for many purposes. Archaeologists doing initial surveys of an area or after collecting data rely on geostatistics to study possible relationships between analyses of terrain use. Chi-square analysis is used as a basic technique to look at the strength of possible relationships, but it is prone to a number of problems arising from autocorrelation.

REVIEW QUESTIONS

1. What is spatial autocorrelation?

2. What is the "ecological fallacy"?

3. Why aren't geographical things and events random?

4. Why is statistical validity important? What are the two types?

5. How are geostatistics used in terrain analysis?

6. What parameters impact chi-square analysis?

7. Explain the modifiable areal unit problem.

8. What is the basic concept behind geostatistical analysis?

9. What is "spatial interpolation"?

10. What do patterns signify about processes in geostatistics?

ANSWERS

1. What is spatial autocorrelation?

Geographic data sets are neither random nor samples; as Waldo Tobler stated, "near things are more related than distant things." Any geographic data will be affected by its location.

2. What is the "ecological fallacy"?

The ecological fallacy refers to the assumption that the statistical relationship observed at one level of aggregation holds at a more detailed level.

3. Why aren't geographical things and events random?

The characteristics and processes of things and events are affected by where they occur.

4. Why is statistical validity important? What are the two types?

Statistical validity measures the reliability of data and analysis. The two types are internal and external validity. Internal validity considers comparisons between measurements made at the same place over time. External validity considers how well data compares to similar measures from other places, other instruments, other types of observations, and other ways of recording data.

5. How are geostatistics used in terrain analysis?

Geostatics can be used in different ways for terrain analysis. A distinction can be made between the initial geostatistical analysis of terrain and analysis that evaluates hypotheses and ideas generated from existing observations.

6. What parameters impact chi-square analysis?

The distance and placement of observations have considerable impacts on chi-square analysis.

7. Explain what MAUP refers to.

The MAUP is a special instance of the ecological fallacy that results when data collected at a more detailed level of aggregation (e.g., census blocks, counties, biotopes) are aggregated to less detailed levels (e.g., census tracts, states, watersheds). Results valid at one level of aggregation may not be valid at another level.

8. What is the basic concept behind geostatistical analysis?

Patterns indicate geographic processes.

9. What is "spatial interpolation"?

It is used for determining characteristics for places or areas based on existing observations.

10. What do patterns signify about processes in geostatistics?

Patterns indicate underlying processes and relationships in geostatistics.

Chapter Readings

Gould, P. R. (1970). Is *Stastix Inferens* the geographical name for a wild goose? *Economic Geography, 46*, 439–448.

Miller, H. (2004). Tobler's First Law and spatial analysis. *Annals of the Association of American Geographers, 94*(2), 284–289.

O'Sullivan, D., & Unwin, D. J. (2010). *Geographic information analysis* (2nd ed.). New York: Wiley.

Tobler, W. R. (1968). Transformations. In J. D. Nystuen (Ed.), *The philosophy of maps* (Discussion Paper No. 12, pp. 2–4). Detroit: Michigan Inter-University Community of Mathematical Geographers, Wayne State University.

Web Resources

○ An illustrated introduction (a pdf file) to the application of basic geostatistical analysis in archaeology can be found online at the *NASA Technical Reports Server*.

○ A GIS-based terrain analysis exercise can be found online at *www.scribd.com/doc/90131023/Exercise-03-Siting-a-fire-tower-in-Nebraska*.

○ For more archaeological examples, see *www.archaeophysics.com*.

○ A tutorial about a real-life adventure in environmental decision making can be found online at *www.nwer.sandia.gov/sample/ftp/tutorial.pdf*.

EXTENDED EXERCISE

Chi-Square Analysis

Objective

Learn basic concepts of chi-square analysis and issues that influence geostatistical analysis.

Description

The chi-square analysis is a straightforward statistical technique to evaluate the validity of hypotheses. While it can get rather complicated for large tables, for simple two-dimensional tables it is still a simple statistical technique. The only thing you should bear in mind when working with chi-square analysis and GI is that GI rarely, if ever, follows a normal distribution.

Instructions

First you need to prepare a combined map of precipitation and forested areas in Isaland (an imaginary island). The null hypothesis is that forested areas are more common in areas with higher rainfall.
Here is a list of the steps:

Prepare combined map.

Indicate precipitation by high/low.

Calculate the ratio of forested land to nonforested land.

Prepare tabulation.

Complete the observed frequencies and expected frequencies matrices.

Determine chi-square statistic.

Calculate Yule's Q using the equation below.

Answer the questions below.

You will use two equations:

Chi-square

$$\chi^2 = \sum \frac{(ObservedFrequency - ExpectedFrequency)^2}{ExpectedFrequency}$$

Yule's Q

$$\frac{ad - bc}{ad + bc}$$

Precipitation

H	H	H	L	L
H	H	L	L	H
H	L	L	H	L
H	L	H	L	H
H	L	H	H	H

Forests

F	—	F	F	F
F	—	—	—	F
F	L	L	F	—
F	—	F	—	F
F	—	F	F	F

Tablulation of
Characteristics
High/Low Precipitation,
Forested/nonforested   Combined

Grid#	Precip.	Forest

Calculation of Forested to Nonforested Land Ratios

Observed frequencies			Expected frequencies		
	Forested	Nonforested		Forested	Nonforested
High precipitation			High precipitation		
Low precipitation			Low precipitation		

Chi-Square Calculations

1	2	3	4	5
Matrix cell	Observed frequencies	Expected frequencies	$(O - E)^2$	$(O - E)^2/E$
a				
b				
c				
d				
				Sum =

To determine the chi-square value we will use an interactive web application or a published chart from your instructor.

Calculate Yule's Q

Use the observed frequencies. A value close to +1 indicates a strong positive relationship between the independent and the dependent variables.

$$((ad - bc) / (ad + bc))$$

Explanation
a = the number of times E1 happened and E2 happened
b = the number of times E1 did not happen and E2 happened
c = the number of times E1 happened and E2 did not happen
d = the number of times E1 did not happen and E2 did not happen

Yule's Q = _____

a, b, c, d stand for the corresponding values from the cells in the observed frequency table

When you are finished, answer these questions.

Questions

1. How strong is the relationship between precipitation and forests in Minnesota?
2. What factors explain the values?
3. What other factors should be considered?
4. What problems do you see in the procedure for this exercise? Are the quadrants a good size for this analysis?

CHAPTER 17

Considering the Past and Future of GIS

Although we usually look to the future now in Western society, apparently the ancient Greeks said we look forward only into the past. In this, the last chapter of this book I've written, that attitude strikes me now as the more appropriate. But I'm writing this for you, the reader, and my perspective may be quite different from yours. You and I indeed may be looking forward to the future, just from different vantage points: I will write accordingly.

In the first edition of *A Primer of GIS*, published in 2008, I began this chapter with these words: "As the computerized collection, processing, and embedding of geographic and cartographic representations increases, GI will become an increasingly integral part of even more activities." In hindsight, that certainly was a modest statement. GIS now seem to many contemporaries to have passed the apogee of their growth and related excitement. For many, they have become only a part (of great importance certainly) of the global information infrastructure. They still play a very important role in the development of sensor webs, also known as the Internet of things, ubiquitous computing, and in some circles the fourth paradigm. And GIS are not fading by any means, filling a vital role when computers and networked data involve location. Arguably, GIS already play the central role as the integrating platform for various technologies and organizations. If it hasn't happened already, GI will be as central as maps were in the last 100 years. Plate 14, an example of an interactive GI application with Google Earth, shows one of many exciting developments.

The rise of GIS to the status of one of the most fundamental information systems in the information society comes because, as the saying goes, it stands on the shoulders of giants. The last three decades have seen tremendous development in geographic information systems (GIS), global positioning systems (GPS), remote sensing technologies, and other key information technologies. The next three decades are likely to witness changes of similar magnitude.

What the future means specifically for the technologies for the use of, and access to, GI is more difficult to say. It seems that GI will move more toward being a commodity that is produced, sold, and exchanged like other information commodities—information about businesses, or information about potential customers, for example. Maybe charges and inequitable access to commercial GI will promote more government support for free or low-cost GI access, kick-starting a new vibrant market that develops applications and supports local democratic decision making. The ubiquity of GI also means increased surveillance by governments and private groups. For many, such unregulated surveillance is a considerable problem for civil liberties; for others, it is a necessary means to assure these liberties. A yet undecided key question in many places is how governments will allow access to their GI. The increasing role of GI in many activities brings both opportunities and challenges.

To begin, and to try to keep our feet on the ground, we should note that GI is still a relatively small player in the information technology (IT) world. U.S. federal spending on IT was about $63.3 billion in 2007 and $74 billion in 2012. To compare, in 2007 the 50 U.S. states spent $23.7 billion on IT. Local government IT spending in the United States in 2007 was somewhat higher: $27.2 billion. On the global scale, the data on just GIS spending alone indicate some interesting trajectories. The entire GIS industry had revenues in 2006, according to Daratech, of $3.6 billion worldwide. By 2012 the worldwide revenue associated with GI had climbed to $15 billion. Although government spending is only part of the spending on GIS, and although comparing spending to revenue is a crude measure, a quick comparison of GIS industry revenue and U.S. total government spending suggests that governmental GIS spending is staying stable, while commercial activities involving GIS have greatly increased. As the numbers suggest, government GIS activities are important, but most government IT funding goes toward other forms of IT. This becomes clear whenever a disaster strikes and the responding agencies are unable to coordinate their GIS because of conflicting standards and organizational responsibilities. Progress is forever being made to address and resolve these issues, but with too many disasters problems impede important life-saving activities.

How government spending will change is a question for fortune tellers. Too many variables influence how government resources are spent. The above considerations lack detailed insights into the changing roles of the GI industry and the government sector. Instead of attempting to prophesize developments, the sections in this chapter focus on some of the questions underlying developments of GI and cartography in the next few years:

- *Where has GIS been?*
- *Where is GIS going?*
- *What are the challenges?*

- *What are the ethical issues?*
- *Who pays for the data?*
- *What are the opportunities?*
- *What is the employment outlook?*

Where Has GIS Been?

Any history of GIS is a partial history. The technologies, institutions, and, most of all, the people are still under examination. We lack information about a wide range of key details. A number of authors (see the Chapter Readings) have offered various studies and stories about how GIS developed. Clearly, the deepest roots of what we call "GIS" today are complex and follow multiple trails. They reach back into developments before and especially during World War II of the first ITs and computing technologies; they fit into a scientific slant in many government institutions to develop rational approaches to decision making and political desires to have empirical approaches to decision making, approaches that meld the strengths of IT with vast repositories of statistical information that governments had begun to collect in the 19th century; they promote a scientific engagement with technologies that provide more robust modes of dealing with information; and they draw on geographers' and cartographers' work attempting to develop visualization techniques. Such techniques would be flexible enough to meet the needs of governments and private industries and were readily taught, at various levels, to the technicians, analysts, and instructors who did that work.

If one were to ask, what is the single most important motivation behind GIS, we might respond that GIS start out with the desire to harness IT for a multitude of needs for representing geographic things and events, primarily with an emphasis on supporting government decision making. Its first proponents and developers came to GI from backgrounds in urban research (Horwood), planning (McHarg), analysis (Fisher), and natural resource management (Tomlinson). Edgar Horwood was interested in examining how urban areas developed and was one of the first individuals to offer an academic course on the processing of GI. In 1963 he founded the Urban and Regional Information Systems Association (URISA), which still exists today. Ian McHarg, a landscape architect, became intrigued by the potential for improving and democratizing planning studies and published his seminal *Design with Nature* in 1968. Howard Fisher had begun working around the same time as Horwood and went to Harvard University in the mid-1960s to develop general-purpose automated mapping and analysis software. Under his direction, the Harvard Laboratory for Computer Graphics and Spatial Analysis became one of the key locations for developing early GIS applications and approaches. In parallel, Roger Tomlinson, who was involved in studies of land use in Canada during the late 1950s through an aerial survey company, began working closely with IBM and developed a system for managing land use information. In 1963 the team working on the project named this the "Canadian Geographical Information System."

Of course, many other researchers were involved. Many other stories are crucial to the myriad ways GIS developed and became such a vibrant and significant technology. Certainly, the many interactions, formal and informal, that took place during the 1950s and 1960s were crucial to the development of GIS as we know it today. Jack Dangermond, a student of urban planning at the University of Minnesota and of landscape architecture at Harvard University, heard about work in Fisher's lab. Grasping the opportunity, he founded a company in Redlands, California, that drew on lessons from the Harvard lab and has since become one of the most successful GIS software and consulting companies, the Environmental Systems Research Institute, or ESRI.

GIS for most of its early years was a domain dominated by large companies and government agencies, especially agencies involved in mapping, such as the U.S. Geological Service (USGS), but the rapid fall of prices for computing hardware in the 1980s and 1990s led to many smaller government agencies and private companies becoming very significant. ESRI has undoubtedly been dominant in many areas, but smaller companies that fill specific niches have become increasingly prevalent since the late 1990s. In the last 15 years, a new group of smaller companies that specialize in the distributed techniques and innovative programming approaches that accompanied the development of the Internet have become increasingly significant. The growing and ever more central importance of standards and specifications indicates that GIS has been changing and continues to change.

Where Is GIS Going?

Although the future will certainly see a continuation of GIS usage for projects involving geospatial distributions and existing application areas, new directions are already evident in the harnessing of GIS operations and data handling to support new applications, especially on mobile devices. If GIS software were a toolbox, an analogy might be just taking the tools for a particular project and greatly improving them by linking them with networked processing and data. Primarily, these new GI functions occur in embedded applications and spatial data infrastructures. You might have used one today to get a navigation map to find your way by car and avoid traffic, locate the nearest public transportation stop, or figure out where to meet friends in the evening. Billions of people around the world now have access to these operations and probably don't ever realize that they rely on GIS. They don't really need to know that GI supports cell phone maps. With the increasing embeddedness of mobile computing through tablets and small, but powerful, laptop computers and sensor networks, it is less likely that people will have occasion or need to understand such things. These developments rest on widely established collections of easily accessible GI in SDI clearinghouses

---

IN DEPTH **What Is a Spatial Data Infrastructure?**

There are many definitions of what makes up a spatial data infrastructure (SDI). Common to all is that an SDI facilitates and helps coordinate the exchange of spatial data. This exchange can be restricted to stakeholders or limited to a defined community. Sometimes the limits are not crystal-clear, but stakeholders and community members find better services, data, and support than others.

Problems arise because any SDI is understood in different ways by its stakeholders and community members. SDIs seem to have been most successful when they support decision making and facilitate interactions between organizations.

(see Chapter 13), and expertise rests with many private and public users of GIS.

Most of us first experience the benefits of infrastructure like this when we use consumer applications. Business-to-business and academic communication may require more robust systems, which may be referred to as location-based services (LBS). These technologies also provide the kind of social networking services used on mobile devices and personal computers. *Location-based services* can be defined as technologies that add geographical functions to other technologies. The most common example is GPS-equipped phones (Figures 17.1 and 17.2). Originally developed to satisfy emergency services' requirements, the addition of GPS to mobile phones has opened up possibilities for cell phone companies and other private entities to develop directory, mapping, and query services for consumers. With online data that is regularly updated, time-based, process data can be visually represented and analyzed. The image in Plate 15 shows the average public bus speeds in the Boston area on one day for a small period of time. Transportation specialists can use such a map to understand or improve services; the green and yellow lines are faster moving, so that can help commuters plan their trip.

In the last 15 years, most LBS applications required programming or licensing software that would be added to mobile phones and other handheld devices. Now, many applications provide a range of functionality to end-users without the requirement of any knowledge of programming. The embedding of LBS with the functionality of GIS in mobile and household devices (e.g., navigation aids that not only find the theoretically shortest route from the store to your home, but also take traffic into account and adjust the route accordingly) is the start for some of an Internet of things that has constantly linked devices, providing helpful services for any need. The basis for many of these developments comes from crowd-sourcing GI and making the information widely and easily available. Much of this is infrastructure, and hidden from consumers, but people working with GI know the value of Open Street Map (OSM) data for providing reliable road, highway, and even path data that is essential to navigation applications. Also known as volunteered geographic information (VGI) it is often combined

FIGURE 17.1. GPS serves a wide range of urban applications.

FIGURE 17.2. GPS also serves a wide range of agricultural and environmental applications.

with official data to create online mapping resources created by nonprofits, news agencies, or even governments themselves to distribute GI to others.

The availability of such services depends on access to GI. In many parts of the world this is an important sticking point. Providing and maintaining the levels of institutional support required for these types of applications is very demanding and very expensive. A spatial data infrastructure (SDI) has become the way of thinking about aligning existing institutions with these developing GIS applications. SDI developments face enormous challenges. Costs are great, retooling existing institutions is always difficult, and finding political support has been challenging. An SDI-based approach seems the best strategy at the moment, but the difficulties have led to the creation of application-oriented data providers that circumvent the complexities of SDI and provide an easy-to-use and consistent system. The development of standards for web mapping have been crucial to these changes.

Indeed, if consumer applications continue to become more important than established SDI, the changes may be so significant that GIS begins to be recognized by other names, or even disappears. What a mobile phone user encounters when using a service to find nearby Italian restaurants has, for that person, little or even nothing to do with GIS. Since GIS may be of great importance only behind the scenes, they may only be relevant to a small group of people involved in the development of services. Increased mobile computing use of GI is certainly bringing about changes in how we think about GIS; the growing number of sensors, semi-autonomous and increasingly autonomous, will bring about yet more changes.

What Are the Challenges?

With changes come challenges. People often struggle to learn to do new things to get tasks done. We have all experienced the struggle involved in replacing the old with the new. The underlying challenges people working with GI face in the coming years are going to lead to GI uses developing in different ways. While there are many, in this section I have picked just the four that I think are most significant to the topics of this book. The first challenge is how people will visualize the growing amounts of GI available. The discussion of Big Data in Chapter 14 focuses on analysis aspects and only touches on visualization issues. Clearly, with the amounts of data now available, we can visualize geographic processes in engaging ways beyond most GIS functionality. Martin Wattenberg and Fernanda Viegas's creation of a dynamic map that shows the wind (*http://hint.fm/projects/wind/*) is a wonderful example of these visualizations' potential. The second challenge is handling and assessing the increasing amount of crowd-sourced data now available. For example, *https://openpaths.cc/* provides a vast repository of maps of trips from place to place, provided by users whose identity has been made anonymous. Many other sites offer this kind of data for particular activities like biking and hiking. How do I know how reliable the data I download from a person for part of Seoul, Korea, is? And how can I assess its accuracy? What do I do if information from a site like this differs from official data?

Most people now can readily access or benefit from others' access to GI. Ideas for ubiquitous computing in which sensors would be placed in all kinds of stationary objects would drastically increase the amount of location data at hand. It's quite unclear how people will make sense of this growing data deluge. The third challenge, and this holds great risks, involves the manner in which corporations and governments will process this location information. Already, some governments seek to ensure that local facilities are presented in online mapping in a positive light. Fourth, and last, the perennial challenge for GI continues to lie in managing error and the limits that arise from making calculations using projected coordinates and scale. Most online maps rely on an adaptation of a Mercator projection (see Chapter 5). The Mercator projection is, however, only accurate along a line of tangency. Away from this line the accuracy of measurements decreases. Scale and generalization impacts the accuracy of measurements as well.

What Are the Ethical Issues?

An important starting point for confronting all challenges related to GI is ethics, or the consideration of the principles of good conduct and how we should act. Ethics is different from morals, which are concerned with right and wrong and good and evil. GIS and mapping lead to many complicated ethical questions. These focus on locational privacy, surveillance, the collection and reuse of data, and the responsibilities of professional groups.

One of the most important ethical issues for the future of GIS is locational privacy. Location-based services open the door to many intriguing and helpful capabilities, but also offer malicious people and groups opportunities to collect information about people's movements and offer businesses possibilities to use the information collected for other purposes. The potential for abuse is so substantial that the infringment on locational privacy has been referred to as "geoslavery" by some well-known GIS figures.

Locational privacy is an issue that dovetails with concerns about increasing surveillance. The exaggerated powers of surveillance technology as shown in popular U.S. movies may be far off, but dense networks of surveillance certainly already exist. No vehicle enters the center of London without its license plate number being recorded. The surveillance networks of communist East Germany showed how much information can be collected even with the help of very modest IT. The increase in the number of surveillance cameras coupled with data from mobile phones and GPS devices and other sensors opens the doors to unimaginable levels of surveillance.

The biggest constraint for surveillance remains the amount of data collected. The collection and reuse of data poses special challenges. Data collected legitimately for one purpose (e.g., the use of prepaid public transportation stored-value cards) can be linked to other data (e.g., images from a surveillance camera) to compile information on individuals that impinges on their privacy. Right now in the United States no blanket law defines privacy and its protections. How legitimate reuse of data should be regulated persists in posing special challenges.

Professional groups recognize a responsibility for establishing principles to help guide their members' activities through these very complex issues. The ethical guidelines or rules presented by various professional agencies provide a useful starting point for considering what is appropriate behavior in a variety of challenging situations.

Educators are also called upon to engage these challenges in their instruction. Generally, ethical issues receive little attention compared to technological issues in technically oriented programs and even in professional programs. The possibility is certainly there to engage ethical issues in conjunction with technical and organizational questions, which seems a more vibrant way to deal with the many changes facing GIS and mapping.

Who Pays for the Data?

Among these changes are substantial institutional changes related to the collection, maintenance, and publication of GI and maps. Large mapping organizations are challenged both internally because of the reduced size of government and externally by the increase in the number of service providers offering GI resources and services to key customer groups. The availability of low-cost GPS equipment has already greatly changed the way that maps for many tourist regions around the world are prepared and sold. Organizations (e.g., *OpenStreetMap.org*) promote the development of freely accessible GI using public domain standards.

One common approach to dealing with these changes has been for organizations investing in GI to attempt to recover all or some of the costs associated with gathering that GI. In some instances, public organizations even attempt to make a profit on the sales of their GI data. While cost recovery has a few successes, for the most part more examples can be found where it has not worked. The reason for the failures has often been that the data is not necessary for the consumer's purposes, he or she could find alternative data sources, or people protested against the perceived overcharging for data.

Technology changes are fueling new approaches to cost recovery that may be more successful. Instead of relatively simple models for charging fees or licenses based on the amount of data requested, service-oriented approaches charge only for the data that has been used. These approaches are certainly the result of distributed geoprocessing, which makes it possible to access only the GI needed for a specific application over the Internet from as many computers, no matter where located, as needed.

Distributed geoprocessing needs special coordination of the underlying infrastructure. Internet connections break frequently, and if for some reason this happens unbeknownst to the computer operator, a serious problem can result. This also applies to the IT used for displaying and working with the GI. Distributed data, as a result, may be incomplete in subtle but important ways. Distributed geoprocessing may not work in many circumstances, no matter the perceived advantages. This is one of the main limitations to these approaches, and as such is very technical in nature, requiring special technical services and/or precautions. Another challenge, along these lines, the reliable provision of Internet-based services over wireless networks and through cables, opens the doors to many opportunities for GIS and mapping.

What Are the Opportunities?

The new and ongoing opportunities in GIS are so many that a separate book on them would be necessary to do them justice. Certainly, advances in technology mean that approaches for mobile computing have received a great amount of attention and will continue to be a focus for developments in the future. The underlying technology requires technically adept individuals; the applications require many more individuals who can make the technology work and best support various consumer and business needs.

Employment

The opportunities for employment in fields working with GI range from research and development to support. Two 2013 reports commissioned by Google evaluating the size of U.S. domestic and global markets for geoservices put the U.S. sector at $73 billion in revenue (Boston Consulting Group, 2012) with half a million jobs and international revenues at $150 to $270 billion (Oxera Consulting, 2013). Correspondingly, the job market is strong. The U.S. Bureau of Labor Statistics indicates that the job outlook from 2010 to 2020 for geographers in the United States will grow by 35%, which is quite a high rate of job growth compared to other "industries." The predictions for cartographers and photogrammetrists specifically is 22% growth, which is also higher than average. Also for surveying and mapping technicians, the outlook is 16% of growth. Across these types of positions you can find a range of activities. On the one end, we find the people involved in creating and developing the technologies and devices. They are often engineers or programmers, but opportunities in this domain also exist for other people who work on testing and refining devices. Support generally involves people who help users work with the technologies and devices as well as through training. In between, many people work on developing and maintaining applications either as programmers or as specialists for particular application domains. Last, but by no means least important, people working in marketing are often involved in assessing consumer demands and helping companies identify profitable areas to work in. It is certainly possible to find individuals who work in all domains during the course of their normal work week—the needs of smaller companies often require such talents.

Opportunities can be found in all sectors. Government positions abound in all areas, but are especially important for applications and to a lesser extent in research and development and in support. Private industry offers a large number of applications as well, and has the most positions in research and development. Nongovernment and nonprofit positions may be scarcer, but many people find that they offer a better balance between quality of life and income than jobs in other sectors.

Summary

GIS has experienced a dynamic past and has adapted to the advent of new technologies. Currently GIS seems set on at least partially morphing into more visualization capabilities focused on specific applications. Support for mobile applications appears to be a key driving force for the next round of GIS development. Opportunities abound in this area, but also continue in traditional GIS areas. These developments create ethical issues. The capability of tracking people through the entire day is now readily possible, but how should we know and control how we are tracked and placed under surveillance? Challenges for the development of GIS touch on these issues, but other fundamental technical and organizational challenges remain.

REVIEW QUESTIONS

1. Do you think that GI will become ubiquitous? Why or why not?

2. How significant are costs for the development of GI?

3. Why is government funding of GIS in the United States so small in comparison to overall funding of IT?

4. GIS's history involves many fields. What are some of the most important fields?

5. What is the name of one of the most important U.S. federal government agencies involved in GIS?

6. What does "LBS" stand for?

7. What is an SDI?

8. What are some important ethical issues for GIS?

9. How are changes to technology impacting cost-recovery models of GI pricing?

10. What is the range of employment opportunities?

ANSWERS

1. Do you think that GI will become ubiquitous? Why or why not?

   There are no right or wrong answers, just thoughtful and unthoughtful ones.

2. In the development of GI, how significant is cost?

   Cost is one of the most significant issues, if not the most significant, no matter what form GI takes.

3. Why is government funding of GIS in the United States so small in comparison to its overall funding of IT?

GIS is only one of the technologies used by government. Major investments in health, welfare, public safety, and many other areas do not involve computing.

4. GIS's history involves many fields. What are some of the most important fields?

Landscape architecture, natural resource management, planning, geography, computer science, and cartography.

5. What is the name of one of the most important U.S. government agencies involved in GIS?

The United States Geological Survey (USGS).

6. What does "LBS" stand for?

Location-based services.

7. What is an SDI?

The alignment of various institutional GIS to support multiple GIS applications and users, some of whom may be unknown and undefined, but important in the future.

8. What are some important ethical issues for GIS?

Some important ones are: What do people do with data? How is the data collected? How will locational privacy be protected? Whether and how should people be charged for the GI they use?

9. How are changes to technology impacting cost-recovery models in GI pricing?

GI is becoming a resource that services draw on. Customers are charged according to their degree of use.

10. What is the range of employment opportunities?

Research, application development, and support.

Chapter Readings

The Boston Consulting Group. (2012). Putting the U.S. geospatial services industry on the map. Retrieved from *www.valueoftheweb.com*.

Chrisman, N. (2006). *Charting the unknown*. Redlands, CA: ESRI Press.

Curry, M. (1998). *Digital places: Living with geographic information technologies*. New York: Routledge.

Fisher, P. F., & Unwin, D. J. (Eds.). (2005). *Re-presenting GIS*. Chichester, UK: Wiley.

Foresman, T. (Ed.). (1998). *The history of geographic information systems: Perspectives from the pioneers*. Upper Saddle River, NJ: Prentice Hall.

Oxera Consulting. (2013). What is the economic impact of Geo services? Retrieved from *www.valueoftheweb.com*.

Pickles, J. (2004). *A history of spaces: Cartographic reason, mapping, and the geocoded world*. New York: Routledge.

Sheppard, E. (1993). Automated geography: What kind of geography for what kind of society? *The Professional Geographer, 45*(4), 457–460.

Web Resources

⟳ OpenStreetMap is a free editable map of the whole world; see *http://openstreetmap.org*.

⟳ The U.S. National Academy of Sciences, Mapping Sciences Committee, periodically publishes reviews and perspectives. The latest is *Beyond Mapping* available online at *http://darwin.nap.edu/execsumm_pdf/11687.pdf*.

⟳ The GISCI website provides a concise list of important principles as well as discussion of ethical issues for GIS professionals. See *www.gisci.org*.

⟳ The Value of the Web site provides reports and infographics on geoservices at *www.valueoftheweb.com*.

⟳ Typical for what many organizations have to do to acquire and maintain GIS, that is, saving and scavenging, see the example at: *www.dailymail.com/story/News/+/2006083137/Frustrated+county+officials+to+use+local+funds+for+mapping*.

⟳ Google Earth is now famous for its mapping and visualization capabilities. See *www.google.com/earth/index.html*.

Glossary

abstraction	Reduces complexity, or simplifies, to highlight essential things, events, and relationships.
associated things and events	Things and events occur together, without any intervening distance.
attributes	Characteristics and qualities of things and events in the representation of data. A database attribute is the particular entry in a field.
band	In remote sensing, band refers to a particular range of wavelength recorded by a sensor.
cadastral maps	Show how land is divided into real property, and sometimes the kinds of built improvements on it.
cadastre	A term that originated in Latin that refers to the registry of land ownership, rights, and responsibilities.
Cartesian coordinate system	A two-dimensional coordinate system.
cartographic representation	The process and choices involved in going from a geographic representation to symbols to communicate with readers.
cartography	The theories, concepts, and skills for describing and visualizing the things and events or patterns and processes from geography and communicating this understanding.

charts	Maps created for use in navigation.
color hue	The result of how different light wavelengths are reflected by surfaces. This is what we commonly call color.
color value	The amount of darkness or lightness of a color.
communication	In which one person's geographical representation is shared with others.
compromise projections	Projections that try to preserve area and sometimes shape.
convention	Uses or procedures agreed upon by experts, but usually that have become common knowledge—for example, that north is the direction oriented at the top of a map.
coordinate system	A grid, that could be used to combine, for example, a projection, a datum, and locational references.
data collection	A necessary step in the process of creating geographic representations of things and events.
data combination	A step in which data with different kinds of measurements are evaluated and perhaps manipulated so that they can be used together; for example, projection, scale, and coordinate system data.
data update	A step in a plan for data management that recognizes that geographical information will change.
database	A collection of data stored in a structured format using a computer. Databases provide the most common computerized means to save and store data for swift access.
database normalization	Making data conform, as well as possible, to relational database principles.
datum	A calibration of location measurements that includes vertical references and is tied to a particular projection.

developable surface	A drawing, or visualization, of a projection in which curvature is represented two-dimensionally.
DGPS	Differential global positioning system.
disciplinary culture	The values, beliefs, and implicit understanding shared in a particular profession or academic discipline.
display scale	The constraints and capabilities of what can be shown on electronic devices, such as a cell phone and an LCD projector.
distortion	The change to an object's actual shape that happens when making a two-dimensional projection of it.
ellipsoid	A shape with three parameters: (1) an equatorial semimajor axis a; (2) a polar semiminor axis b; (3) and a flattening factor f.
EMR	Electromagnetic radiation.
event	What happens at a specific moment in a process.
field	A subdivision in the data in a "record."
fitness for use	A determination of how well a map serves its intended purpose.
framework	A normative and acceptable system for showing things, events, space, relationships, and associations.
generalization	The abstraction of features to reduce complexity in maps and GI.
geographic information science	(GISci), the field concerned with the underlying theories and concepts of GI.
geographic representation	How aspects of the world are shown on a map or as GI.
geography	Human and environmental phenomena and processes that take place on the earth's surface.
geoid	The most accurate model of the earth's surface.

georelational model	A common way to store vector GI that relies on topology, tables, nodes, and chains.
geospatial	How objects and activities are arranged, with some reference to a worldwide, or global, scale.
GNSS	Global navigation satellite system.
GPS	Global positioning system.
great circle distance	The shortest route between two points on the earth.
ground truthing	People on the earth's surface make measurements or observations that are matched with those made by remote sensing devices. This calibrates the remote sensing data and may be used to correct remotely sensed data in situations where there will be no ground observations.
indigenous culture	The beliefs and knowledge of the people who lived in an area before it was colonized, as those beliefs and knowledge have been passed down to their descendants.
latitude	The distance measure, in degrees, of a point on the earth's suface from the equator.
longitude	The distance measure, in degrees, of a point on the earth from a fixed meridian, which has been the Greenwich meridian for a long time now.
map	A form of output that can be produced from a geographic information system.
mental map	How a person visualizes geographic and spatial relationships, and a drawn representation of it.
modeling	Using GIS data to create new visualizations or predict how processes, land surfaces, and so on will change and develop over time.
MPLIS	The Multipurpose Land Information System (MPLIS) is perhaps the best known and most significant attempt at sharing administrative data. GI is stored using a common coordinate system.

multipath error	Positional errors created by GPS satellite signals that reflect differently from ground, vegetation, buildings, and so on.
national culture	The values, beliefs, and implicit understanding commonly held by the majority of people living in a particular nation.
orthophoto	Image or data in which elevation change or elevation differences have been removed; also called planimetric.
PDOP	Positional dilution of precision.
principle	Standard procedure that people in a field follow— for example, when a cartographer chooses a projection to make a map.
projection	The process in which a three-dimensional shape (the round earth) is shown two-dimensionally on a flat map.
projection aspect	A projection's orientation to the earth's surface.
projection property	Whether a projection represents angles, areas, or distances (from one or two points) as they are found on the earth's surface. Projections cannot retain both angles and areas.
propaganda	Information manipulated to fit particular ideological, political, or social goals with a malicious intent.
purpose	What GI or a map is intended to communicate.
quality	A map's fitness for its intended uses.
raster	A form of geographic representation in which space is fixed and the amount or degree of an attribute in that space is measured.
related things and events	The things and events are connected in terms of distance.
reliability	The degree to which geographic information or a map fulfills what it was designed for, that is, its purpose.

remote sensing	The collection of data without directly measuring it.
representation	Characterizing geographic phenomena in a way that gives it meaning.
scale	The relationship between the size of a representation and the actual area of what is being represented.
SDI	Spacial data infrastructure.
semiotics	The study of signs.
spatial	How objects are arranged, with reference to location, extension, movement, and so on.
spatial resolution	The size of a unit as recognized by a sensor.
spectral resolution	The range of wavelengths a sensor can record.
spheroid	A simple round model representing the shape of the earth.
supervised classification	Pixels are assigned to categories.
symbol	A picture or other graphic pattern that stands for something a map user can recognize.
table	The part of a relational database that is stored in a file of its own and that is "related" by the database to other tables.
temporal resolution	Indicates how often a satellite passes over and/or takes readings of the same place.
tesselation	A systematic subdivision of a space using tiles of one or more geometric shapes.
thematic map	A map that shows specific topics and their geographic relationships and distributions.
thing	An aspect of the world that is static.
topographic map	A map that shows the physical characteristics of the area depicted.

transverse projection	Turned 90° from the usual orientation of the projection; for example, the Transverse Mercator projection.
tuple	Represents a single data item in a table.
unsupervised classification	Pixels are assigned to categories by software, according to values, but without instruction from the operator.
vector	A form of geographic representation in which an attribute is fixed while the space it occupies is measured—the geospatial extent of features and objects. Common vector geometries are points, lines, and areas in GIS software.
visualization	Depiction of phenomena. Computers allow dynamic visualizations that add a temporal component.
WAAS	Wide area augmentation system.

Sources and Credits

FIGURE 6.12. From *https://zulu.ssc.nasa.gov/mrsid/docs/gc1990-utm_zones_on_worldmap.gif.*

FIGURE 6.14. From *www.spatial-effects.com.* Reprinted by permission of Geoff Dutton.

FIGURE 7.7. Courtesy of Michael Lutz.

FIGURE 8.1. Photo courtesy of Crain Inc.

FIGURE 8.2. From *http://erg.usgs.gov/isb/pubs/booklets/topo/topo.html.*

FIGURE 8.6. From Campbell, J. B., and Wynne, R. H. (2011). *Introduction to remote sensing* (5th ed.). New York: Guilford Press. p. 393. Reprinted by permission of Guilford Publications.

FIGURE 8.7. From *http://commons.wikimedia.org/wiki/Image:Global_Positioning_System_satellite.jpg.*

FIGURE 9.3. From *http://lmynasadata.larc.nasa.gov/images/EM_Spectrum3-new.jpg.*

FIGURE 9.4. From *http://rst.gsfc.nasa.gov/Intro/Part2_5.html.*

FIGURE 9.5. From *http://landsat.gsfc.nasa.gov/education/compositor/.*

FIGURE 9.6. From *http://www.csc.noaa.gov/products/sccoasts/html/rsdetail.htm.*

FIGURE 9.7. From *http://earthasart.gsfc.nasa.gov/images/netherla_hires.jpg.*

FIGURE 9.8. From *http://earthobservatory.nasa.gov/Newsroom/BlueMarble/.*

FIGURE 11.1. From MacEachren, A. M. (1994). *Some truth with maps: A primer on symbolization and design.* Washington, DC: American Association of Geographers. Reprinted by permission of Alan M. MacEachren.

FIGURE 11.2. Images from *http://erg.usgs.gov/isb/pubs/factsheets/fs01502.html.*

FIGURE 11.7. From *http://egsc.usgs.gov/isb//pubs/gis_poster/.*

FIGURE 11.17. Above, from *www.epa.gov/ada/about/thumbnails/gis.htm.*

FIGURE 11.20. From Cromley, E. K., & McLafferty, S. L. (2011). *GIS and public health* (2nd ed.). New York: Guilford Press. p. 121. Reprinted by permission of Guilford Publications.

FIGURE 11.21. From Cromley, E. K., & McLafferty, S. L. (2011). *GIS and public health* (2nd ed.). New York: Guilford Press. p. 120. Reprinted by permission of Guilford Publications.

FIGURE 12.1. From Department of Anthropology, Smithsonian Institution, catalogue No. 398227, photo by P. E. Hurlbert.

FIGURE 12.2. Courtesy of Mona Domosh and the Auburn University Archives, Auburn, Alabama. Thanks to Auburn University Libraries Special Collections and Archives.

FIGURE 12.4. Courtesy of Jonathan Cinamon.

FIGURE 12.6. From Kosonen, K. (2000). *Kartta ja kansakunta: Suomalainen lehdistökartografia sort-ovuosien protesteista Suur-Suomen kuviin, 1899-1942.* Helsinki: Suomalaisen Kirjallisuuden Seura. Reprinted by permission of Katariina Kosonen.

FIGURE 13.1. Courtesy of Hennepin County Library Special Collections.

FIGURE 13.2. Courtesy of MetroGIS, St. Paul, Minnesota.

FIGURE 13.4. From *http://www.loc.gov/item/2011586687/.*

FIGURE 14.1. Courtesy of *Metrogis.org.*

FIGURE 14.2. Courtesy of the Minnesota Department of Natural Resources (DNR). © (2015) State of Minnesota, Department of Natural Resources.

FIGURE 14.4. Courtesy of *infoamazonia.org.*

FIGURE 15.1. From *http://tx.usgs.gov/geography/austgrth_large.htm.*

FIGURE 15.3. From Dangermond, J. (1979). A case study of the Zulia Regional Planning Study, describing work completed. In G. Dutton (Ed.), *Urban, regional and state applications* (Vol. 3, pp. 35-62). Cambridge, MA: Harvard University Press. Adapted by permission.

FIGURE 15.7. From *www.csc.noaa.gov/benthic/mapping/applying/pdf/bmdredge.pdf*.

FIGURE 15.8. From Chrisman, N. R. (1999). What does "GIS" mean? *Transactions in GIS, 3*(2), 175-186. Adapted by permission of Nicholas R. Chrisman.

FIGURE 15.9. From *http://pubs.usgs.gov/mf/2000/mf-2346/mf-2346so.pdf*.

FIGURE 16.1. From *www.nwer.sandia.gov/sample/ftp/tutorial.pdf*.

FIGURE 16.2. From *www.nationalatlas.gov/natlas/Natlasstart.asp*.

FIGURE 16.3. From *www-personal.umich.edu/~mejn/cartograms*.

FIGURE 16.4. Courtesy of Jason Menard.

FIGURE 17.1. Image courtesy of Magellan Navigation Inc.

FIGURE 17.2. Image courtesy of Magellan Navigation Inc.

PLATE 1. From *www.dot.state.mn.us/tmc/trafficinfo/map/refreshmap.html*. Reprinted by permission of Mn/DOT.

PLATE 2. From *http://regions.noaa.gov/great-lakes/wp-content/unloads/2013/06/wrecks-of-1913-map. png*.

PLATE 3. From *www.jpl.nasa.gov/spaceimages/details.php?id=PIA12133*. Courtesy NASA/JPL-Caltech.

PLATE 4. From MacEachren, A. M. (1994). *Some truth with maps: A primer on symbolization and design*. Washington, DC: American Association of Geographers. Adapted by permission of Alan M. MacEachren.

PLATE 6. From *http://science.nature.nps.gov/im/units/ucbn/educ_camas/gis.cfm*.

PLATE 7. From *www.training.fema.gov/emiweb/downloads/DF109%20GPS%20IG.pdf*.

PLATE 8. From *http://earthasart.gsfc.nasa.gov/images/garden_hires.jpg*.

PLATE 9. Reprinted by permission of Transport for London.

PLATE 11. From *http://geography.wr.usgs.gov/science/dasymetric.html*.

PLATE 12. Courtesy of Chris Lloyd.

PLATE 13. Courtesy of Chris Lloyd.

PLATE 14. Image courtesy of Jason Dykes. Reprinted by permission of Google Earth™ mapping service.

PLATE 15. Reprinted by permission of Andy Woodruff.

Index

Note: *f* or *t* following a page number indicates a figure or a table.

About the Author

Francis Harvey, PhD, is Head of the Department of Cartography and Visual Communication at the Leibniz Institute for Regional Geography and Professor of Visual Communication in Geography at Leipzig University in Germany. He was formerly Associate Professor in the Department of Geography, Environment, and Society at the University of Minnesota. Dr. Harvey's research addresses a range of central issues in geographic information science, including visualization, ethics, values, institutional aspects, cadastral issues, pedagogy, and overlay algorithms. His teaching has covered the use of evolving geographic information technologies for undergraduate and graduate students in the United States and internationally. Currently, he is Chair of the International Geographical Union Commission on Geographical Information Science.